SpringerBriefs in Sociology

SpringerBriefs in Sociology are concise summaries of cutting-edge research and practical applications across the field of sociology. These compact monographs are refereed by and under the editorial supervision of scholars in Sociology or cognate fields. Volumes are 50 to 125 pages (approximately 20,000–70,000 words), with a clear focus. The series covers a range of content from professional to academic such as snapshots of hot and/or emerging topics, in-depth case studies, and timely reports of state-of-the art analytical techniques. The scope of the series spans the entire field of Sociology, with a view to significantly advance research. The character of the series is international and multi-disciplinary and will include research areas such as: health, medical, intervention studies, cross-cultural studies, race/class/gender, children, youth, education, work and organizational issues, relationships, religion, ageing, violence, inequality, critical theory, culture, political sociology, social psychology, and so on. Volumes in the series may analyze past, present and/or future trends, as well as their determinants and consequences. Both solicited and unsolicited manuscripts are considered for publication in this series. SpringerBriefs in Sociology will be of interest to a wide range of individuals, including sociologists, psychologists, economists, philosophers, health researchers, as well as practitioners across the social sciences. Briefs will be published as part of Springer's eBook collection, with millions of users worldwide. In addition, Briefs will be available for individual print and electronic purchase. Briefs are characterized by fast, global electronic dissemination, standard publishing contracts, easy-to-use manuscript preparation and formatting guidelines, and expedited production schedules. We aim for publication 8–12 weeks after acceptance.

Christophe Guibert • Bertrand Réau
Editors

Employment and Tourism

New Research Perspectives in the Social
Sciences

 Springer

Editors
Christophe Guibert
ESTHUA, Tourism and Culture
University of Angers
Angers, France

Bertrand Réau
1LAB40
LISE-CNRS-CNAM
Paris, France

ISSN 2212-6368   ISSN 2212-6376   (electronic)
SpringerBriefs in Sociology
ISBN 978-3-031-31661-6   ISBN 978-3-031-31659-3   (eBook)
https://doi.org/10.1007/978-3-031-31659-3

This Springer imprint is published by the registered company Springer Nature Switzerland AG
The registered company address is: Gewerbestrasse 11, 6330 Cham, Switzerland

Acknowledgments

We are grateful to the Tourism Research and Higher Education Association (AsTRES) for supporting the English translation costs of this research and to the ESTHUA Faculty of Tourism, Culture and Hospitality (ESTHUA) of Angers University (France) for funding the editorial costs, thus making our research accessible to a wider public.

Furthermore, we are also appreciative to the National Conservatory of Arts and Crafts (CNAM) for holding a seminar on the theme of labour and employment in tourism in January 2019, and to the Tourism Studies GIS (GIS études touristiques) for funding the 2020/2021 Tourism Colloquium, where the exchanges enriched the reflection on the theme of tourism employment, and this study was one of the outcomes of both seminars.

We would also like to thank Mr. Jean-Yves BART, the translator of this book, Ms. CHEN Sijie, the doctoral candidate in charge of editing, and the staff at Springer for their help and support, in making the English version of the study available.

Acknowledgments

Contents

Editors and Contributors

About the Editors

Christophe Guibert is a sociologist, professor at the University of Angers (ESTHUA, Faculty of Tourism, Culture and Hospitality), and researcher at the "Spaces and societies" laboratory (UMR CNRS 6590). For the past 20 years, he has been examining multiple dimensions attached to tourism practices (public policies, jobs, social and cultural uses, gender, etc.) in France, but also in various foreign countries (China, Taiwan, Morocco, USA, etc.). His work is part of a dispositional-ist and multi-methodological sociology. He has managed research contracts and published numerous scientific articles relating to these themes. He has published or been editor of *L'univers du surf et stratégies politiques en Aquitaine* (2006, L'Harmattan), *Tourisme et sciences sociales* (2017, L'Harmattan), *Les "sports de nature": une catégorie de l'action politique en question* (2017, Éditions du Croquant), *Emplois sportifs et saisonnalités* (2011, Logiques sociales, L'Harmattan), and *Les mondes du surf, Transformations historiques, trajectoires sociales, bifurcations technologiques* (2020, MSHA). Since 2016, he has managed two licenses and a master's degree in the field of coastal tourism in Les Sables d'Olonne, a delocalized branch of the University of Angers (France).

Bertrand Réau is a Professor at the Cnam, entitled to direct research, and holds the "Tourism and leisure travel" Chair. His recent work focuses on tourism practices and the social uses of time, the challenges of the globalization of science and disciplinary recompositions around studies, the relationship between tourism and ethnicity in Southeast Asia, and the development of theme parks around the world. He is notably co-author of *Sociologie du tourisme* (2016), *La sociologie de Charles Wright Mills* (2014), and *Researching Elites and Power* (2020, Springer) and author of *Les Français et les vacances. Sociologie de l'offre et des pratiques de loisirs* (2011).

Contributors

Aurélie Condevaux is assistant professor at IREST (Institute for Research and Higher Studies on Tourism) – Paris 1 Panthéon-Sorbonne and member of EIREST (Interdisciplinary Research Group on Tourism). Her main research interests lie in cultural performances and tourism in New Zealand and Tonga, the political aspects of heritage processes – with a particular focus on Intangible Cultural Heritage – in the Pacific and, more recently, the sharing economy and virtual tourism.

Etienne Guillaud is PhD in sociology and is currently a post-doctoral fellow for the CLASMER project, at UBO in Brest (France). He defended a PhD on professional wear process among nautical sports teachers, at the University of Nantes in 2018. His research focuses on working and employment conditions in the leisure, sport and tourism sectors.

Sébastien Jacquot was elected director of IREST (Institute for Research and Higher Studies in Tourism) at the University of Paris 1 Panthéon-Sorbonne, by the IREST Board, on July 8, 2020.

Teacher-researcher (lecturer) in geography at the University of Paris 1 Panthéon Sorbonne at IREST, since 2009. His teaching focuses on research methodology, follow-up of field workshops, courses on tourist development, heritage, tourist imaginaries, metropolization and tourism, comparative urban policies, etc. Since 2010, he has been coordinating the master's in tourism – Development and Territorial Tourism Planning. He has also been director of IREST since 2020, after 3 years as deputy director. As a researcher, he is a member of EA EIREST and an associate member of UMR PRODIG. He conducts research on heritage issues, concerning both evident heritage assets (World Heritage) and what can be characterized as infra-heritage, questioning the boundaries of heritage. His work also focuses on tourism, in its links to the territorial fabric (tourist metropolization), to memory tourism, to the digital humanities via the study of tourist social networks.

Francis Lebon is professor in education sciences at the Université Paris Cité, member of the CERLIS, CNRS (UMR 8070). His research focuses on popular education professionals and the division of labour in primary schools. He is currently conducting two surveys: one on the policies and practices of access to law; the other on teaching practices in the first grade.

François Mandin is the director of the Center for Maritime and Oceanic Law (CDMO). HDR lecturer, Sciences and techniques of physical and sports activities, Nantes University.

Gabriele Pinna is researcher at the University of Cagliari in Sardinia (Italy). Gabriele Pinna received his PhD in sociology at the University of Paris 8. His main research interests are work and employment in tourism and vocational education and training (VET). He has published articles in French, Italian and English in various international scientific journals and a book in French entitled *Travailler dans l'hôtellerie de luxe. Une enquête ethnographique à Paris* (2018).

Gérard Rimbert, Doctor in sociology, Gérard Rimbert has been an associate lecturer at the Cnam since the end of 2020.

Chapter 1
Introduction. What Does Working in the Tourism Sector Mean?

Christophe Guibert and Bertrand Réau

Abstract Touristic practices and destinations are becoming increasingly widespread in more and more places and countries. Recent years have also witnessed a diversification of the ways of doing tourism. Still, at each level of the "tourism system" (Équipe MIT (2005) Tourismes. 2, Moments de lieu. Paris: Belin) – a "system of actors, practices, and places" (Knafou R, Stock M (2003) 'Tourisme', In: Dans Levy J, Lussault M. Dictionnaire de géographie et des sciences de l'espace et du social. Paris: Belin), the functions performed by professionals, although they are changing, are crucial to deliver tourism services that meet the expectations and demands of tourists. As a market-oriented activity, tourism entails, from the step of booking (a trip, a hotel room, a paragliding class, for instance) to the performance of the service (e.g. a visit at a heritage site or a winery, a sporting activity), a wide range of embedded professional activities requiring sector-specific hard and soft skills. While the tasks and functions of tourism workers are varied within this "plural space" (Guillaud E (2018) De l'attrait à l'usure: Les trajectoires professionnelles des éducateurs sportifs en nautisme, Thèse de Doctorat en sociologie, Université de Nantes, Nantes, France, p.43), pay and skill levels can also be wildly different within the same country or from one country to the next.

Keywords Social sciences · Tourism · Employment · State · France

Touristic practices and destinations are becoming increasingly widespread in more and more places and countries. Recent years have also witnessed a diversification of the ways of doing tourism. Still, at each level of the "tourism system" (Équipe MIT, 2005) – a "system of actors, practices, and places" (Knafou & Stock, 2003), the

C. Guibert (✉)
University of Angers, Angers, France
e-mail: christophe.guibert@univ-angers.fr

B. Réau
LISE-CNRS-CNAM, Paris, France

© The Author(s), under exclusive license to Springer Nature
Switzerland AG 2023
C. Guibert, B. Réau (eds.), *Employment and Tourism*, SpringerBriefs in
Sociology, https://doi.org/10.1007/978-3-031-31659-3_1

functions performed by professionals, although they are changing, are crucial to deliver tourism services that meet the expectations and demands of tourists. As a market-oriented activity, tourism entails, from the step of booking (a trip, a hotel room, a paragliding class, for instance) to the performance of the service (e.g. a visit at a heritage site or a winery, a sporting activity), a wide range of embedded professional activities requiring sector-specific hard and soft skills. While the tasks and functions of tourism workers are varied within this "plural space" (Guillaud, 2018, p.43), pay and skill levels can also be wildly different within the same country or from one country to the next.

It is in fact classically in macroeconomic terms that national ministries dedicated to tourism assess the attractiveness of destinations: economically and socially, tourism is a major sector of the national economy in numerous countries including France, Morocco, Tunisia and Spain. France, for instance, is "the world's first tourism destination", with nearly 90 million foreign visitors hosted every year. "Tourism constitutes a key sector of its economy: it amounts to nearly 8 per cent of the GDP, 56.2 billion euros in revenue, and two million direct and indirect jobs" (French assistant ministry in charge of tourism 2020). However, a more refined analysis of tourism jobs is needed to move beyond this very general, quantitative approach. This is the ambition of this book, whose contributions, mostly original, are based on theoretical frameworks and empirical material mainly from France.

The economic, technological, cultural and social transformations that can be observed in most professional sectors have also been impacting tourism. While tourism jobs tend to be associated with hospitality staffing/receptionist jobs (at tourist offices, hotels, campsites, resorts, heritage sites, etc.) and service jobs (sports instructors – skiing, surfing, sailing, etc. –, tour guides, camp coordinators, etc.), many professions are physically and symbolically far removed from direct contact with tourists. Service staff (Pinna, 2013) such as maids or dishwashers in the restaurant industry, in particular, although they are "invisible to the world" (Castel, 1995), hidden from tourists or limited to short, rare interactions with them, do play crucial, necessary roles in the smooth running of a tourism operation.

Tourism jobs are usually structured by the length of local tourism seasons. On most European coastlines, for instance (in Spain, France, Italy, but also in the UK, Belgium, etc.), service jobs in tourism are for the most part offered as fixed-term summer contracts, whose duration rarely exceeds three to four months, between May and September. "Without seasonal workers, tourism is not possible", as Dethyre (2007) has pointed out. These seasonal workers may be hired in the food industry (bars, restaurants), in the hospitality sector (hotels, campsites) or in recreational services (in specialized associations and companies). They might also work as lifeguards, as vendors on stalls or in waterfront shops, etc. In France, their working contracts differ from the standard fixed-term (CDD) contracts according to the Ministry of Labour: "Seasonal work is characterized by the execution of tasks that are normally meant to be repeated every year at more or less fixed dates, depending on the seasonal cycle (harvest, picking, etc.) and on collective lifestyles (tourism). This variation in activity must be outside the employer's control". The definition of the duration of "the season", the period during which seasonal workers can be hired, has significant implications for employers and the professional branches of the tourism sector. Indeed, at the end of a seasonal contract, unlike for the standard fixed-term contracts, the employer is under no obligation to provide termination benefits equivalent to 10 per cent of the worker's total gross remuneration; it is therefore in the employer's interest that the tourism season is as long as possible. The difficulties in recruiting seasonal workers following the COVID-19 lockdowns have evidenced a malaise in the sector and the declining appeal of these jobs.

Owing to the characteristics of this singular professional sector whose legal boundaries are quite blurry, a wide variety of social backgrounds can be found among seasonal and year-round tourism workers. From students in summer jobs to salaried workers with multiple jobs who have worked "seasons" for years, to youth in job integration programmes, living in a touristic destination, who take advantage of the local economic manna for one or two summers or workers who combine summer and winter seasons or switch between them, social properties (age, gender, diplomas and skills, work experience, etc.) are quite heterogeneous. This has been shown for instance by Aurélien Gentil, who evidenced three categories: the "transients", the "regulars" and the "locals" (Gentil, 2013). According to Seuret (2007, p.46), in an article for the French magazine *Alternatives économiques*, some of the specificities of the seasonal status are the lack of *prime de précarité* (an end-of-contract allowance) unlike in CDD contracts, the rarity of labour inspections, sometimes illegally long work weeks, and a tendency towards unchecked flexibility. The authors of the report for the Fondation Abbé Pierre (2003), who refer to seasonal workers as the "drudges of tourism", have made similar observations: "The picture is gloomy and difficulties are not limited to the question of housing: there is also precarity, demanding hours, low wages, breaches of labour laws". These grey areas and the inherent volatility of seasonal work have been denounced for years: "a little-known sector is most often a badly treated sector", as the Le Pors report (1998) put it.

The holidays are characterized by a strong spatial and temporal concentration (respectively in mountain or coastal resorts and in the summer and winter). The fact is that being an employee of (or freelancing for) a firm or an association providing tourism services means working out of time and out of place – in other words outside of the places and temporalities that define the ordinary living settings of tourists (Guibert, 2012; Sebileau, 2014). Still, 'the standard employment model (meaning salaried work, under an open-ended [CDI], full-time contract, part of a complete, uninterrupted career), often heralded as the pillar of the "wage-earning society" (Castel, 1995) (Grégoire & Join-Lambert, 2017, p.5) remains the professional norm. In France, a 1982 law cemented this by legally defining the open-ended CDI contract as the yardstick for employment. According to the INSEE's 2018 national employment survey, "Wage earners amount to 88.3 per cent of people employed in France in 2018. Among them, 84.7 per cent work under an open-ended contract (CDI) or are civil servants, 10.5 per cent under a fixed-term contract (CDD), 3.0 per cent as temps, and 1.8 per cent are in training." Sarfati and Vivés (2018), p.122) found a preference for the CDI within the space of possible professional statuses among their interviewees: "Regardless of their professional trajectories, age, gender, etc. [...] the CDI is the most desirable contract and the one that offers the best working conditions. [...] To justify the desirable character of the CDI, two main reasons have been advanced: the guaranteed pay and increased opportunities of access to housing".

The "insurance societies" described and analysed by the sociologist Robert Castel (2003) embed individual trajectories within collective social protection schemes. This is the case of the open-ended contract in France, and more generally of the status of full-time year-round salaried worker, which offers a substantial protection and a "social citizenship", as Castel terms it (Castel, 2013, p.28). In other words, the work provides economic, social but also symbolic guarantees. According to Méda (2018, p.5), work is "a foundational component of the social order, it largely determines the place of individuals in society. [...]. Working is a norm." A factor of production, a source of social integration, allowing for the distribution of income and benefits, work is the result of a combination of historical processes. Tourism jobs contribute to the de-standardization of work: as quite a few of them are characterized by variable durations, fixed-end contracts, non-standard working hours, part-time work, etc., they put the workers in question outside of the professional settings that are usually perceived to be the most stable and profitable.

For some of the seasonal workers, who operate in this manner year in and year out, trajectories become erratic, fragmented, atypical, non-linear and geographically mobile. In a number of studies, some of the seasonal workers interviewed expressed an intention not to pursue any further in that line of work, confirming Christian Baudelot's analysis: the more an occupation is difficult, badly paid and temporary, the more work (in the broader sense) is perceived as fundamental to self-fulfilment and to the search for happiness (Baudelot et al., 2003). The workers' choices of trajectories that lead them to alternate between periods of work (summer) and of non-work are in fact often at odds with the advice and recommendations given by their relatives. As a result of a long-term political effort to "attach strong protections to labour" (Castel, 2013, p.28), families consider stable wage work to be the desirable norm, and in contrast, the seasonal worker is identified as "the symbolic face of precarious work" (Dethyre, 2007). There is, however, more pressure of that sort among the seasonal workers whose parents enjoy(ed) a stable social and economic situation (teachers, retailers, business owners, etc.): "While the fear of falling has spread to the entire body social, it has acquired a special significance in the middle classes" (Ferreira, 2006, p.172). The perceived risk of generational social downward mobility and/or the low yield of the degrees obtained —in short, the "fear of decline"—are the main factors of parental reluctance towards the idea of seasonal, temporary tourism work.

Ultimately, studying touristic employment through the lens of its structural specificities calls for reconsidering multiple dimensions (economic, cultural, temporal, geographical, etc.), which reflects the scientific dynamism of a perpetually changing research field. The economic, technological, cultural and social transformations that can be identified in most professional sectors also affect the tourism field, in the following ways, for instance:

- The rise of more or less formalized procedures resulting from the development of the Internet, such as the online booking of tourism products (trips, tours, accommodations, etc.) impacts the core jobs in the tourism sector. New jobs, new professional tasks are emerging. How is the deployment of new technologies in strategic digital plans accepted and internalized by staff at a grassroots level? How do these workers perceive the evolution of their functions, of their jobs (Sotiriadis & Varvaressos, 2016)? How have tour operators made social changes (such as downsizing the workforce in travel agencies)? What positioning/repositioning strategies have companies in the sector resorted to in the face of internationalized competition? What form does the tourism supply take when individuals no longer play a role in the services provided (as in online tourism services such as podcast guided tours, e-concierges, etc.)?
- While tourism jobs tend to be associated with hospitality and service, many professions are physically and symbolically far removed from direct contact with tourists. More and more tourism-related jobs are back office positions, but they add considerable value to the service. What forms of recognition can front desk and back office workers accordingly hope for?
- As noted above, tourism jobs are characterized by specific geographical and temporal dimensions. This raises a number of questions: how can one work during the school holidays and have a family life? How can one maintain a steady relationship when working the "seasons" (in the summer at the beach as a sailing instructor, in the winter as a mountain guide)? What collective solutions (such as employers' groups in some regions) are being devised to facilitate employment for seasonal workers and keep them coming back? What results have those achieved? In what ways can the sector overcome its temporal constraints to make tourism work in general more appealing?
- On a related note, many studies have shown that commitment to work out of passion characterizes some occupational sectors; tourism is one of them. Whether they are passionate about historic buildings, surf (Guibert & Slimani, 2011) or horse riding, tourism professionals embark on careers that are inspired by their own personal practices and interests. How do these workers negotiate the place of their passion in their job? Considering what these jobs entail behind the scenes, can we speak of a sense of disenchantment or disillusionment among these workers? Are these tourism jobs part of a career plan or merely stepping stones on the way to finding more promising work?
- Beyond the possibility of combining passion and work, what is the nature of the employment relationship of tourism workers, who are actors in the sector by choice or by necessity? What are its specificities, what forms does it take depending on the skill levels required? How can it be approached within the framework of a renewed social exchange theory (Cropanzano et al., 2017) to help anticipate behaviours at work instead of merely observing them and to elicit desirable behaviours for tourism sector organizations that are facing new demands (lifestyle hotels, narrative approach, enriched personal interactions, etc.)? How do the sets of positive and negative behaviours of tourism workers, who deal with varied and sometimes unpredictable experiences at work, play out in practice

(engagement/withdrawal, good citizen behaviour/counterproductive behaviour/
trust/defiance/ surface/deep emotional labour, etc.)? What role do the social partners, whose involvement is often minimal, play in defining social contracts?
– Considering the cyclical effects of the economic crisis that has been hitting
  European and North American countries in a variety of ways, how are human
  resources policy strategies being defined or redefined? How are corporate HR
  strategies implemented in the field? The global growth of tourism puts pressure
  on the tourism supply which in turn leads to more demand for manpower. This
  workforce is also courted by other sectors that often offer more appealing working conditions. What new strategies do tourism businesses use to access new
  pools of workers and keep them coming back? Are the trainings offered (on site
  or in e-learning form) adjusted to the jobs and the skills sought after by tourism
  business? What role are soft skills given in training programmes and hiring
  policies?

While the health and economic crisis has had structural effects on tourism, it is difficult to obtain a precise statistical overview of these impacts, as the outlines of the
sector are fuzzy. It is indeed complex to tell apart touristic uses from everyday uses
in culture, sport, leisure activities, food, etc. The only solution left is to rely on surveys of professionals. A quantitative survey conducted during the first French
Covid-19 lockdown (March–July 2020) with 702 tourism professionals yielded a
snapshot of experiences and representations in the sector at that particular time
(Guibert & Réau, 2021). Overall, the impact of the public health crisis and of the
economic fallout from the lockdown was almost exclusively considered as "catastrophic" (60%) or "very negative" (28%). Workers in the private and cooperative
sector were particularly concerned, with a "catastrophic" situation for 70% of businesses.[1] Regarding the travel sector (travel agencies, tour operators, etc.), a survey
conducted by two specialized labour unions (Entreprises du Voyage and Syndicat
des Entreprises du Tour Operating) between 8 and 20 October 2020 found that six
out of ten travel firms considered laying off salaried workers.[2] Forty percent of the
421 respondents planned to cut at least forty percent of their jobs. This announced
wave of redundancies was partly motivated by cash flow issues, as tour sales fell by
seventy-five percent in October 2020, for instance. In another sub-sector, the union
of independent hotel and restaurant workers sounded the alarm in mid-September
2020, noting that nearly 47,000 jobs had been cut in hotels in the first semester of
2020 and expecting that 30,000 more might be destroyed by the end of that year.[3]

---

[1] Source:« L'impact de la crise sanitaire sur les compétences et la formation professionnelle sur le
secteur du tourisme et des loisirs », Programme de recherche pour la DARES, Ministère du Travail
(2020–2021).

[2] « 6 entreprises du voyage sur 10 comptent licencier », *L'Echo Touristique*, 30 octobre 2020.

[3] « L'hôtellerie française est en train de vivre le plus grand plan social de son histoire » », *L'Echo
Touristique*, 16 septembre 2020.

It clearly transpires from the above that implementing a social science approach to the study of employment and work in tourism requires taking into account varied social phenomena and complex mechanisms. The considerable heterogeneity of the professional situations under study gives a wide scope and a dynamism to a research field that is primed for more investigations and international comparisons. This book is intended to review existing knowledge on the sector mainly in sociology, law and anthropological, drawing on a selection of original empirical research on the one hand and more theoretical contributions on the other.

Thus, the first three chapters, by Christophe Guibert, Gabriele Pinna and Etienne Guillaud, examine, from a sociological point of view, the specificities of the professional careers of employees in the coastal tourist economy, respectively in the department of Vendée (north-west) in France, in Sardinia and then more globally on the Atlantic coast. These three texts analyse the social mechanisms according to which employees of the tourist economy continue in the profession, the links between jobs and professional statuses, and the question of the recognition of the work of supervising nautical activities within the coastal territories. The chapter proposed by François Mandin questions, from the point of view of law, the links between employment, tourism and sport. What are the regulations - and the effects of these regulations - that govern the professions of "action sports" instructors in France? Francis Lebon's chapter focuses, from the point of view of the sociology of education, on the notion of "social tourism": if the range of jobs in this specific type of tourism is wide, the economic precariousness seems to be equally important. Aurélie Condevaux and Sébastien Jacquot, respectively an anthropologist and a geographer, examine, in their joint chapter, the professional changes that characterise the tourism sector. In particular, they question the place of salaried employment and entrepreneurship in the field of Internet reservation platforms. In his chapter, Bertrand Réau studies recruitment methods in a multinational company in the tourism sector. It appears that recruitment methods are a function of the dispositions and social properties of candidates and recruiters. The recruitment situation is thus a privileged moment to analyse, from a sociological point of view, the connivances of habitus. The text proposed by Gérard Rimbert follows a series of investigations on the effects of the health crisis on the tourist economy. The author notes a massive and brutal increase in the economic situations suddenly entering into a relationship of dependence of the professionals towards the public support devices, in particular of the French State. The conclusion of the book is based, finally, on an original collective research work. The scientific preoccupations of social science researchers have, in fact, been largely reactivated thanks to the effects of the Covid-19 crisis on the tourist economy. On the French scale, a survey was carried out for the Ministry of Labour in 2021 in order to capture the reality of tourism employment and the possibilities offered in the field of professional training. The purpose of this conclusion is to present the main original results.

As mentioned earlier in this introduction, one of the scientific objectives of this book is to introduce French-speaking research - and therefore currently translated - to an English-speaking readership of English, American or Australian researchers and academics (or other countries, of course). The international academic community should indeed be able to access research work initially conceived in languages other than English, particularly in the light of a research theme - tourism employment - which is of growing scientific interest. The scientific risk undoubtedly concerns the excessive autonomy of research work, which is still too often compartmentalised and closed in on itself, whether in terms of the theoretical frameworks mobilised or the empirical fields. Indeed, approached from a contemporary perspective, this book aims to limit a scientific deficiency: that the labour market outside the English-speaking world has not yet been sufficiently covered and studied. May the chapters and analyses proposed in this book contribute modestly to rebalancing this observation!

References

Baudelot, C., Gollac, M., & Bessière, C. (2003). *Travailler pour être heureux ? Le bonheur et le travail en France*. Fayard.

Castel, R. (1995). *Les métamorphoses de la question sociale*. Folio Essais.

Castel, R. (2003). *L'insécurité sociale. Qu'est-ce qu'être protégé*. Seuil.

Castel, R. (2013). Individus, risques et supports collectifs. *Idées Economiques et Sociales., 171*, 24–32. https://doi.org/10.3917/idee.171.0024

Cropanzano, R., Anthony, E. L., Daniels, S. R., & Hall, A. V. (2017). Social exchange theory: A critical review with theoretical remedies. *Academy of Management Annals, 11*(1), 479–516.

Dethyre, R. (2007). *Avec les saisonniers. Une expérience de transformation du travail dans le tourisme social*. La Dispute.

Équipe MIT. (2005). *Tourismes. 2, Moments de lieu*. Belin.

Ferreira, C. (2006). La 'crainte de la chute': Le retour des classes moyennes dans l'analyse du politique. *Mouvements, 3*(45–46), 168–174.

Gentil, A. (2013). Entre ancrages temporaires et mobilités saisonnières: l'installation permanente des travailleurs mobiles du tourisme sur le littoral atlantique. *Espace Populations Sociétés, 2013/1-2*, 111–124.

Grégoire, M., & Join-Lambert, O. (2017). Marges de l'emploi et protection sociale. Une analyse sociohistorique. *Travail et Emploi, 149*, 5–16.

Guibert, C. (2012). Les effets de la saisonnalité touristique sur l'emploi des moniteurs de sports nautiques dans le département des Landes. *Norois. Environnement, Aménagement, Société, 223*, 77–92.

Guibert, C., & Réau, B. (2021). Les travailleurs du tourisme dans la tourmente. *L'Économie Politique, 3*, 36–46.

Guibert, C., & Slimani, H. (2011). *Emplois sportifs et saisonnalités. L'économie des activités nautiques : enjeux de cohésion sociale*. L'Harmattan.

Guillaud, E. (2018). *De l'attrait à l'usure: Les trajectoires professionnelles des éducateurs sportifs en nautisme*. Université de Nantes, Nantes, France.

Knafou, R., & Stock, M. (2003). Tourisme. In J. Dans Levy & M. Lussault (Eds.), *Dictionnaire de géographie et des sciences de l'espace et du social*. Belin.

Méda, D. (2018). *Le travail*. PUF.

Pinna, G. (2013). Vendre du luxe au rabais: Une étude de cas dans l'hôtellerie haut de gamme à Paris. *Travail et Emploi, 136*, 21–34.

Sarfati, F., & Vivés, C. (2018). De l'intérim au CDI intérimaire. Se stabiliser dans le salariat pour limiter la subordination. *Sociétés Contemporaines, 2*, 119–141.

Sebileau, A. (2014). Rester après la saison : l'économie symbolique du néoruralisme balnéaire. *Juristourisme, 163*, 31–34.

Seuret, F. (2007). Les saisonniers, des salariés au rabais. *Alternatives Économiques, 264*(12), 46–46.

Sotiriadis, M., & Varvaressos, S. (2016). *Crucial role and contribution of human resources in the context of tourism experiences: Need for experiential intelligence and skills*. Emerald Group Publishing Limited.

Christophe Guibert is a sociologist, professor at the University of Angers (ESTHUA, Faculty of Tourism, Culture and Hospitality), and researcher at the "Spaces and societies" laboratory (UMR CNRS 6590). For the past 20 years, he has been examining multiple dimensions attached to tourism practices (public policies, jobs, social and cultural uses, gender, etc.) in France, but also in various foreign countries (China, Taiwan, Morocco, USA, etc.). His work is part of a dispositionalist and multi-methodological sociology. He has managed research contracts and published numerous scientific articles relating to these themes. He has published or been editor of *L'univers du surf et stratégies politiques en Aquitaine* (2006, L'Harmattan), *Tourisme et sciences sociales* (2017, L'Harmattan), *Les "sports de nature": une catégorie de l'action politique en question* (2017, Éditions du Croquant), *Emplois sportifs et saisonnalités* (2011, Logiques sociales, L'Harmattan), and *Les mondes du surf, Transformations historiques, trajectoires sociales, bifurcations technologiques* (2020, MSHA). Since 2016, he has managed two licenses and a master's degree in the field of coastal tourism in Les Sables d'Olonne, a delocalized branch of the University of Angers (France).

Bertrand Réau is a Professor at the Cnam, entitled to direct research, and holds the "Tourism and leisure travel" Chair. His recent work focuses on tourism practices and the social uses of time, the challenges of the globalization of science and disciplinary recompositions around studies, the relationship between tourism and ethnicity in Southeast Asia, and the development of theme parks around the world. He is notably co-author of *Sociologie du tourisme* (2016), *La sociologie de Charles Wright Mills* (2014), and *Researching Elites and Power* (2020, Springer) and author of *Les Français et les vacances. Sociologie de l'offre et des pratiques de loisirs* (2011).

Chapter 2
Making It Through the Tourist Season: Summer Work in Seaside Resorts

Christophe Guibert

Abstract The tourism professions in coastal areas are, in France, structured by the duration of the tourist seasonality of the territories. Faced with various social norms to contract a job on an open-ended contract, what "keeps" - or not - employees whose trades are characterized by contracts fixed-term, sometimes very dense hours and a significant turnover? In support of a qualitative survey conducted in Vendée, the results allow us to specify the specifics of seasonal work, to draw up differents social conditions and to explain the representations that seasonal workers make of their respective situations.

Keywords Seaside resorts · Working conditions · Seasonal worker · France

Tourism sector jobs on the French coastlines are classically dependent on the duration of local tourist seasons. On the Atlantic coast, for instance, a majority of service jobs in the tourism sector is characterized by fixed-term contracts in the summer, whose term rarely exceeds three to four months, between May and September. While tasks and functions vary within a 'plural space' (Guillaud, 2018), wages are rather disparate (from minimum wage to over double minimum wage, including tips and other bonuses) and high skill levels or degrees are not required. Employees working under a *contrat à durée déterminée* (CDD, fixed-term contract) and 'seasonal' workers can thus be hired in food services (bars, restaurants) as waiters, dishwashers, etc., in the hotel industry (hotels, campsites) as receptionists, maintenance staff, etc., in recreational services (in specialized associations and companies) as customer service agents, activity leaders, state-registered sports instructors, etc. they may also work as lifeguards (at a pool, a lake or on coastal beaches), vendors in seaside shops, etc. The social properties (age, sex, diploma(s) and skills, professional experience(s), etc.) of these workers vary widely, from students on a summer job to employees having held multiple seasonal jobs for years, and from young

C. Guibert (✉)
University of Angers, Angers, France
e-mail: christophe.guibert@univ-angers.fr

people seeking to enter the job market to locals taking advantage of their home-town's economic opportunities for a summer or two, or to salaried workers doing summer and winter seasons every year.

This distinctive sector is however characterized by a fuzzy legal status (Seuret, 2007). According to the Ministry of Labour, a seasonal worker's contract should be distinguished from a traditional CDD fixed-term contract: 'Seasonal work is char-acterized by the performance of tasks that are usually expected to be repeated every year, on more or less unchanging dates, depending on the rhythms of the seasons (harvesting, picking...) or of collective lifestyles (tourism...). This variation in activity must be beyond the employer's control.'[1] The definition of the duration of 'the season', which is effectively the period during which seasonal workers can be hired, is an important matter for employers and the respective branches of the tour-ism sector. Indeed, it is in the best interest of these employers for the tourist season to be as long as possible: unlike regular fixed-term contracts, seasonal contracts are not subjected to mandatory termination benefits (*prime de précarité*) amounting to 10 per cent of the employee's total gross remuneration.

In light of this, and considering that in France securing a CDI (open-ended con-tract) is seen as a must by a wide variety of (cultural, union, political, etc.) actors,[2] how do these workers endure these jobs characterized by fixed-term contracts, sometimes very long hours, in which 'you don't count your hours' and the 'pace of work is hectic', as some interviewees told me, and working conditions are often demanding? How do they, so to speak, make it through the season? What are the effects of the structure of touristic employment on living conditions, especially when it comes to family life as workers age? What type of gains and benefits do these short-term salaried workers hope to register? Do only economic interests count, or might there also be a symbolic dimension specific to these jobs? What meanings do these workers ascribe to their jobs? Could passion be a driving force explaining the commitment of some of them? Are these jobs part of a professional career plan, or are they considered as obligatory steps towards finding a more stable position? All these questions have been under-researched in social science when it comes to tourism jobs, with the exception of a few French-language studies pub-lished in recent years (see in particular: Réau, 2006; Dethyre, 2007; Guibert & Slimani, 2011; Rech & Paget, 2012; Guibert, 2012; Pinna, 2013; Gentil, 2013; Sebileau, 2014; Hoibian, 2014; Guillaud, 2017).

For the purposes of this research, a qualitative study was conducted on the Atlantic coast in the Vendée department, for the most part in the resort of Les Sables d'Olonne, complete with nearly 60 semi-directive interviews (autumn 2017 and

---

[1] Source: https://travail-emploi.gouv.fr/droit-du-travail/les-contrats-de-travail/article/le-travail-saisonnier. See also Article L.1242–12 of the *Code du travail* [French labour code].

[2] Per the French labour code, 'The open-ended employment contract (CDI) is the normal and gen-eral form of the employment relationship. By definition, it has no fixed term. It may be terminated by unilateral decision of either the employer (dismissal on personal or economic grounds, retire-ment) or the employee (resignation, retirement), or for reasons outside the parties' control (e.g., force majeure).'

autumn 2018) with waiters in bars and restaurants, maintenance and reception staff in tourist accommodations, surfing and sailing instructors, and lifeguards. Interviewees also differed with respect to age and experiences of seasonal work, sex, skills and geographic origin. Questions pertained to the meaning they ascribed to their work, their life stories, past experiences and professional expectations – trajectories –, their (material, familial, etc.) living conditions and their social properties. The results provide us with an overview of the range of possible experiences of seasonal work, allowing us to identify its specificities and to get a clearer picture of subjective representations of this work. This chapter is an invitation to consider their annually renewed professional experiences and activities in a contextualized manner, at the intersection of multiple dimensions (professional, familial, and residential) of social life.

2.1 Tourism Sector Jobs: Some Contextual Data

According to the French ministry of tourism, as of 2020: 'Tourism is a key sector of [the] economy: it accounts for nearly eight per cent of GDP, 56.2 billion euros in revenue and two million of direct and indirect jobs'. Very small businesses (VSB) weigh heavily in the sector: nine out of ten firms have fewer than ten salaried employees. VSBs in the tourism sector employ mostly seasonal workers and have few or no year-round employees to train and manage work teams. While the sector has a high potential for integration into the job market, the level of training required for most jobs is fairly low, with 80% of entry-level positions. This type of salaried work is largely associated with economic precarity and with the absence of prospects of long-term employment or career development. The high evaporation and turnover rates make it difficult to retain employees. The high proportions of fixed-term or part-time contracts and the limited career prospects lead the most qualified workers to exit the sector and opt for other ones. Tourism employment is characterized by often demanding working conditions, as well as low skill levels and a limited integration on the job market due to the domination of fixed-term contracts.

At the scale of the Vendée department, tourism has a strong economic and social impact. According to a 2017 survey by Vendée Expansion (an offshoot of the local chamber of commerce and industry), it accounts for 2.1 billion euros in revenues in 2016, including 288 million in the hospitality industry and 190 million in the food industry.[3] Each year, millions of tourists (36.3 million overnight stays in 2017) come to Vendée, mostly in July and August (which make up for 61 per cent of yearly visitors). Sixty-three per cent of tourism jobs in 2016 were temporary, meaning operating under fixed-term CDD contracts – i.e., 23,398 out of 37,018 salaried jobs. Among these temporary jobs, nearly 4600 employment contracts are for durations

---

[3] Source: Vendée Expansion, Service d'Observation et d'Information économiques, 'Le poids économique du tourisme en Vendée', 2016, 6p.

of one to two months over the summer period. Two thirds of tourism jobs in Vendée are located on the coast. Seventy-four percent of seasonal employees are aged 18 to 25; single and childless employees are over-represented.

Beyond this overview of the socioeconomic situation of these workers, a finer, qualitative analysis is needed to yield more detailed insights into what being a seasonal worker is, how these jobs are experienced and perceived by the workers themselves. While pursuing an economic interest, committing to the work out of 'passion', expanding one's social network and acquiring symbolic prestige are the leading reasons brought forward by the seasonal workers I asked about their professional choices, the fact remains that their trajectories unfold in alternative temporalities (during most people's holidays) and localities (holidaying spots). Being a summer tourism worker on the coast means signing a short-term contract, going back and forth between work and non-work periods, and being subjected to (or choosing) structuring forms of geographical mobility. This creates a deviation from social standards in which the CDI, open-ended contract is held up as the rule, the 'norm of employment' in the words of Dominique Méda (2018), and professional, geographic and domestic stability (being in a committed relationship, having children) are socially valued. How do seasonal workers experience these unorthodox professional postures, 'non-aligned' in Erving Goffman's (2002) metaphorical expression about gendered ranks and alignments in schoolyards which effectively reinforce the 'social insecurity' analyzed by Robert Castel (2003)?

2.2   Out of Time and out of Place

2.2.1   Stable Employment and the 'Wage-Earning Society' as Protective Norms

The holiday season is characterized by a strong spatial (resorts) and temporal (summer) concentration. The fact it is that being an employee of (or freelancing for) a firm or an association providing tourism services means working out of time and out of place – in other words outside of the places and temporalities that define the ordinary living settings of tourists (Sebileau, 2014). Still, 'the standard employment model (meaning salaried work, under an open-ended [CDI], full-time contract, part of a complete, uninterrupted career), often heralded as the pillar of the "wage-earning society"(Castel, 1995)' (Grégoire & Join-Lambert, 2017) remains the professional norm. A 1982 law cemented this by legally defining the open-ended CDI contract as the yardstick for employment. According to the INSEE's[4] 2018 national employment survey, 'Wage earners amount to 88.3 per cent of persons employed in France in 2018. Among them, 84.7 per cent work under an open-ended contract (CDI) or are civil servants, 10.5 per cent under a fixed-term contract (CDD), 3.0 per

---

[4] French National Institute of Statistics and Economic Studies.

cent as temps, and 1.8 per cent are in training.'[5] Young people aged 15 to 29 are statistically more often employed under temporary contracts than the population at large: 'Securing a CDI appears, particularly for that population of youth, as a necessary condition for a stable professional life, as well as achieving social inclusion (especially in terms of access to housing)' (Portela & Signoretto, 2017). The preference for the CDI within the space of possible professional statuses is thus mainly functionalist: 'To justify the desirable character of the CDI, two main reasons have been advanced: the guaranteed pay and increased opportunities of access to housing' (Sarfati & Vivés, 2018).

Conceptions of the welfare state pertaining to solidarity, control of the future and social progress, which consist in 'anticipating upward social mobility trajectories' (Castel, 2003) are difficult to reconcile with the status of seasonal worker in coastal tourist resorts like Les Sables d'Olonne. On the contrary, in terms of organization of production, seasonal work requires a 'de-standardization[6] of work' that relies on the flexibility of contract terms and working hours, as well as on the variation of tasks and professional settings. For some of the seasonal workers, who operate in this manner year in and year out, trajectories become erratic, fragmented, atypical, non-linear and geographically mobile. In my qualitative research, I also found that the choices of these workers, consisting in alternating periods of activity (in the summer) and non-activity, run counter to the advice and recommendations dispensed in their immediate social surroundings. The parents of the seasonal workers I interviewed are generally those who most readily assert the importance of having a stable, full-time, year-round job, with an open-ended contract. Resulting from a long-term political effort to 'attach strong protections to work' (Castel, 2003), these parental representations internalize the pre-eminence of salaried work, contrasted with the seasonal worker 'as the emblematic figure of the precarious worker' (Dethyre, 2007). This pressure is greater among the population of seasonal workers whose parents have or had a stable social and economic position (teachers, retailers, managers, etc.): 'While fear of falling has spread to the body social at large, it takes on a meaning of its own in the middle classes' (Ferreira, 2006). The perceived risk of social downclassing and/or the low profitability of one's diplomas – in short, the 'fear of decline' – are what cause the most intense parental reluctances as to the seasonal worker status. Many examples of such fears can be found in the qualitative empirical material I have collected, as attested by the following situations:

François is 32 years old and comes from the northern city of Arras. He graduated high school with a STT/ACC diploma (commerce/retail) and subsequently obtained

---

[5] INSEE, 'Une photographie du marché du travail en 2018'. Last accessed 14 July 2020, https://www.insee.fr/fr/statistiques/3741241

[6] Contrasting with this process of 'de-standardization' of labour in the broader sense, most seasonal jobs are actually involve 'hyper-standardized' duties. Waiting tables, doing maintenance in tourist accommodations and sites, informing customers about the variety of services available to them are low-level professional activities that are characterized by recurrence and sometimes constant repetition. Likewise, a surfing or sailing instructor introducing tourist customers to the activity will also have to repeat gestures and tasks extensively.

a two-year technician certificate (BTS) in 'Negotiation and digitalization of customer relations'. Additionally, he was trained in surfing instruction at the UCPA
Bombannes centre,[7] in Carcans, near Bordeaux. He has been a surfing instructor in
Vendée for seven years, after 'selling fruit and veg on markets in Jard-sur-Mer quite
a few times and doing catering for two winter seasons, we'd work in November and
December for PASO in Vendée'. His parents, both schoolteachers, took his choice
of not pursuing a career path related to his studies 'very badly'. Being in a lower
social position than his relatives, he has had to contend with the criticisms of his
parents (he quoted them on numerous occasions during the interview), especially
his mother, who perceives education as the foundation of professional success, having herself failed at achieving an entirely fulfilling career through education.

> They'd wave a carrot in front of me, they kept going 'if you get this grade, we'll get you
> this, we'll get you that', so I'd go 'yeah, alright', I'd just do the bare minimum I needed to
> do, but it's a good thing they were there to give me a bit of a boost, otherwise I would…
> Well, you can't rewrite the past, but maybe a manual job would've suited me better from the
> start, something tangible.
> They took it very badly. My mother wasn't able to do the studies she wanted to, because
> of her family, things were pretty tough with her parents. And clearly they never wanted to
> pay for her studies, even though she'd graduated high school with great grades. She wasn't
> able to do the studies she wanted to, and so she wanted to pay me the studies I wanted to do:
> 'If you want to do medicine, we'll pay for the studies'. They wanted me to have a job, to go
> as far as I could. You know, that's a sentence that stays with you: 'Study – the more you
> study, the more you get to be able to choose'.
> 'A surfing instructor? But what do you do the rest of the time, how does that work?
> You're not making enough money…' Yeah, but the thing is, I have a fulfilling job. But that's
> hard for them to hear. And my mother obviously tells me I've wasted my potential for study-
> ing: 'You could have studied, have had a career, be a banker'…

Marie is 23 years old and holds a master's degree in communication. Previously, she
worked the winter season in ski resorts and the summer season on the Vendée coastline. She has now been employed under a fixed-term CDD contract by a recreation
centre in Notre-Dame de Monts for two years. She says she is paid 'slightly over
minimum wage'. According to her, her parents, both teachers in the north-western
city of Angers, did not perceive her choice of pursuing seasonal work well:

> My father was a bit reluctant since I've studied, right, I mean I have degrees, so he would've
> liked me to go for jobs that were more related to my studies, considering I initially got a
> master's in communication. So my father sort of wondered why I'd done all that if I was
> going to end up doing quote-unquote seasonal work, but over time, and with my job at the
> seaside centre, he's come to realize that it's a job like any other.[8]

Chloé is 21 years old and holds a vocational high school (*baccalauréat*) degree in
sales, after an apprenticeship in the hospitality sector (equivalent to a CAP[9]) in a

---

[7] The national union of outdoor sport centres (UCPA) is a French association created in 1995 to
promote outdoor sports.

[8] Interview, October 2017, Les Sables d'Olonne.

[9] The certificate of professional aptitude (CAP) is a low-level, secondary school French vocational
qualification.

school in Vendée's administrative capital, La Roche-sur-Yon. Her father manages a sports gear shop and her mother is a saleswoman. She has worked as a waiter for three years, exclusively in the summer, in a 'mid-range' restaurant at the Les Sables d'Olonne harbour. Her future is elusive, but securing an open-ended CDI contract remains her ambition:

> So I'm trying to find a restaurant here, I'd like to get a CDI, to be hired with a CDI. Then again I can always try retail or give England a try, go to England for a few months [...]. You know, CDD [contracts] are fine when you're young, but they're not great, you can't be sure about anything! Yeah, I'd love a CDI, for sure… but in the food industry, it's not easy finding them. [Ultimately, she expresses agreement with her parents' idea that being a waiter with a CDD is only a temporary situation:] Oh yeah, they don't think about it in the same way, sure, they've always told me: 'Well, I don't think you're going to… you know, you're living day-to-day at the moment, but you're not going to be doing this still later, right'![10]

Quentin is 29 years old and graduated high school with a science diploma, after which he stopped studying 'to work, when I was eighteen'. He works as a receptionist at a hotel six months a year, having alternated between summer and winter seasons for eight years now. The occupations of his parents (her father is an electrician and his mother a social worker) indicate a declining social standing, both economically and in terms of status. He intends to stop doing seasonal work and look for a more stable situation, allowing him to envision a family life:

> Since I'm getting older, every year I lose a mate or two, seasonal workers who leave and go on to do something else. I've been telling myself it's my turn, you know? And I'm thinking, too, I'm 29 now. It's a serious thing, quitting. The seasons are really very nice. But I'm starting to think about mundane stuff, like having a wife, kids, and this makes it really complicated. [Since he committed to this lifestyle, he's 'managed on his own'] When I was 18, and I got my high school degree, my parents told me: 'Okay, if you're going to study, we'll pay for that, we'll cover rent, etc. but if you want to go and do something else, in that case, you'll have to manage by yourself.' I didn't really want to study, I didn't know what I wanted to do. So it hasn't always been easy with my parents, they didn't really get it, I suppose.[11]

Alexia is 22 years old, the daughter of a salesman and of a nursing assistant at the Saint-Gilles-Croix-de-Vie hospital. She holds a technician certificate (BTS) in tourism. She's been working seasons for three years, first as a sales assistant and then as a receptionist. Having just earned her degree, she is not considering a career as a seasonal worker, but looking for a CDI contract related to her area of training, which she sees as an evident condition to be able 'to make plans' for the future:

> As far as what I want, I'd like to find a stable, CDI job in reception work, it suits me very well right now. [...] Well, in the sense that it's an obligation, if I want to buy a house I'll need a CDI, if I want to buy a car too, otherwise I won't be able to get a loan, they [banks] will turn down my application: you have to have a CDI if you're looking to make plans in life.[12]

---

[10] Interview, October 2017, Les Sables d'Olonne.

[11] Interview, October 2018, Les Sables d'Olonne.

[12] Interview, October 2018, Les Sables d'Olonne.

Social pressure, which is expressed in the form of a norm that should be followed and of family incentives, is not the only hindrance to pursuing a long-term career in seasonal work. How, indeed, can one juggle summer work during the holidays and family life? How can one sustain a long-term relationship when working 'the seasons' (summers on the beach as a sailing instructor and winters as a mountain guide, for instance)? How does ageing impact the workers' own perceptions of seasonal work?

2.2.2   Age Effect, Family Life and the Structure of Seasonal Work: 'It's not compatible'

Structurally, being a seasonal worker on the coast of Vendée involves working during the summer season, when tourists are most numerous. While some positions span longer periods – roughly April to October – most contracts are short, July–August affairs. Periods of work coincide with the summer school holidays, and in some cases the spring and autumn school holidays. There are few jobs to be found locally in the off season. Surveys conducted by the local branch of the unemployment agency Pôle Emploi[13] show that 'seasonality is extremely prevalent on our territory: in Les Sables d'Olonne, 57% of projected hires are seasonal! This results in a fairly high unemployment rate, because there are fewer jobs in the winter'.[14] This in turn raises the problem of the complementarity between summer and winter jobs for seaside resorts like Les Sables d'Olonne.

In addition to this local specificity, as workers age, they may find it difficult to continue or to picture themselves sticking with a career in seasonal work, characterized by alternating summer and winter seasons: 'It is a fact that for all seasonal workers, when you grow older, you've got a family, kids, it takes its toll. A career change is often a solution for these people.'[15] The manager of a five-star campsite in the hinterland of Vendée has also noticed this. She employs 80 seasonal workers every year, including 50 per cent on two-month contracts, 25 per cent on three to four-month contracts, and 25 per cent on six-month contracts. She says 'there's a pressure to get a CDI, to have a kid, to own a property. And those who make a career of seasonal work, at some point they get tired: you're always over-pressured, so yeah, some of them quit.'[16] Indeed, many seasonal workers do report that they do not see themselves alternating summer/winter, work/non-work seasons indefinitely. Beyond the incentives and interests in conceiving the CDI contract as socially and

---

[13] Pôle Emploi is a public administrative institution in charge of employment in France. Its main core tasks are to provide return-to-work assistance, grant unemployment benefits, and connecting business and jobseekers.

[14] Interview, 6 February 2019.

[15] Interview, 6 February 2019.

[16] Interview, 28 January 2019.

economically more stable, the reasons for this are to be found in the domestic sphere, and in the work's toll on 'family life'. In other words, 'reconciling family and work'(Trancart et al., 2009; Périvier & Silvera, 2010) is the issue here. The vital need for economic profitability of businesses in the summer, the lack of work schedule arrangements for childcare, the relative absence of commitment and measures from firms in the sector to limit domestic constraints, etc. are problems that are raised quite acutely and directly for seasonal workers. Out of step with the 'typical' temporalities of work, the work schedules of these workers also 'makes them out of step with some temporalities of family life, in which weekends, bank holidays and school holidays may be considered as important moments' (Guillaud, 2017). They clash with the paces of family life, affecting their relationship with their partner or the children's visits. Seasonal workers can only rarely take days off at the same time as their children, whose holidays tend to coincide with peak tourist seasons.

However, field data nuance this seemingly unequivocal idea that only work has an impact on domestic life. Some personal and organizational arrangements suggest that we cannot think of family-work reconciliation only in terms of the regulation of domestic time by work, following a one-way normative model, from the work sphere to the domestic sphere. Indeed, while my qualitative analysis shows that seasonal work does have structural effects on families, family organization also has effects on the work in return. This more nuanced articulation of social temporalities, in a continuum of sorts, reflects the idea that for some workers, constraints and advantages are equally found in the work schedule and in the family schedule. Several seasonal workers report using their off-season time off to 'travel', 'spend time with the family', 'get some rest'. Others mobilize resources that enable them to make it through the intense summer season unscathed, like 43-year-old David, who asks his parents to handle childcare when he works at a surfing school. He is married with three children and his parents were teachers. He manages a surfing school that is open only for the touristic season, seven months a year. As his partner works with him, managing this reversed temporality (working in the summer, being inactive in the winter) is unproblematic for his domestic life. While being 'independent and having ownership of his work instrument' (Méda) allows David to more readily accept demanding working conditions, negotiations and arrangements with his own parents facilitate childcare during rush periods at work. His relationship to work has changed now that he manages his own business and has children: 'My wife works with me at the surfing school, and my three young kids are on the beach all summer long. Sometimes the grandparents come and give us a hand. In July, August, it does get a little intense sometimes as you get tired and you have these long working days, but that's only for two months, we take it easy for the rest of the year'.

Recurrences in the discourses of interviewed seasonal workers do suggest that domestic time and work time are difficult to reconcile. Seasonal workers with several years of experience are not the only ones who think that combining atypical work schedules and a love life, a marital life or building a home with children is a challenge. For many of the workers I interviewed, both women and men, these two words are 'closed off' (Bozon, 2009), and juggling them is 'not easy', 'compli-cated', or even 'incompatible', as in the cases recounted below.

Twenty-one-year-old Chloé, who I quoted earlier, is well aware of the challenges caused by the seasonal worker status: 'You work at lunchtime and then in the eve-ning, so in the afternoon you can't necessarily make appointments, and even later, for those who have kids and all that, it's not easy!'.

Nicolas, a single, childless 26-year-old man who lives at his parents' place in Les Sables d'Olonne, holds a vocational degree (*baccalauréat*) in business. His father is a civil servant and his mother is a seamstress. He works as a waiter in the food industry during the summer, in various restaurants on the harbour or on the Remblai waterfront promenade of Les Sables d'Olonne. According to him, the pace of work during the summer season gets 'intense': he claims to have 'no social life' then, 'except with my co-workers', 'nada'. The words used by Nicolas to refer to this total absence of social life does not so much reflect a sense of exclusion as the pro-cess of 'social isolation' described by Robert Castel (1995). Resulting in weak social networks, seasonal work does not fully play a role of 'social integration' for Nicolas.

Louis is 20-year old son and lives with his parents, a deep-sea fisherman and a waitress in a retirement home. After graduating high school with a literature diploma, he did not go to college. He has been working summers as a multi-skilled employee in a grill house on fixed-term CDD contracts since the age of seventeen. He is single and childless. Stability, his 'comfort zone', and his plans to 'resume higher education in September, to be a sound engineer' steer him away from alter-nating working seasons:

> I know there are people, even here at Les Sables, who do summer and winter seasons, and they go abroad. My best friend does that, she does six months of work, and six months of travelling. They do what they want. It's their choice. It's a lifestyle. Depends what you're looking for. Someone who loves traveling will make it so that they have a job that allows them to travel whenever they want. It's not for me, I like having my comfort zone, my little routines and rituals (laughs). I don't mind leaving it behind but knowing you have a place to call home is a comfort. Having something stable. People who do this are around thirty, tops, but the more they'll grow old, the more they'll want something else. I know few sea-sonal workers who are 40 or 50.

Arthur, is 30 years old, single and childless, and holds a technician certificate (BTS) in hotel and restaurant work. His father is a farmer and his mother cooks for local public facilities. From June to October, he works as a head waiter in a restaurant located on the harbour of Les Sables d'Olonne, under a fixed-term (CDD) contract. He says he is paid '1600€ after tax, plus roughly 600€ in tips every month'. He discusses variations in needs and desires along with age:

> It's tougher to have a family life when you're doing this job. Working weekends, for instance, if you have kids, isn't great. You don't get to enjoy them. It's fine when you're young. I got a divorce this year, and one of the main reasons is that I wanted to work seasons again. Which means working late, starting early, being tired... My partner didn't want me to. [...] Seasons lend themselves to flings, but if one of the partners doesn't work like that, it gets tough. Because of distance, the working hours...

Julie is an 18-year-old student in Nantes, the regional capital. Her parents manage campsites: her father's is in the small rural town of Landevieille, a few miles from Les Sables d'Olonne, and her mother's is in Olonne sur Mer. She started out 'lending a hand' in the family business when she was fourteen. Her recent seasonal work has involved 'reception work, activity monitoring, a little bit of cleaning'. She is in a stable relationship with a boyfriend she met during the summer season:

> We met this summer at the campsite, I was doing activities and he was doing reception. He wants to go on working seasons but my plans are different. At the moment we're both living at my place in Nantes, and staying with our parents in the weekends. He wants to find work for the winter season, he's looking for a job, but since I'm a student I'm going to stay in Nantes. So for the time being our plans are different. You can't predict the future [laughs], I suppose we'll see how the relationship is going in a few months!

Théo is 20 years old and comes from a working-class family of Noirmoutier. His father is a market gardener (he grows potatoes) and his mother is a childminder. He holds a technical certificate (BTS) in tourism and works summer seasons on campsites. He is childless and claims to be single by choice, to remain geographically mobile and able to quench his thirst for new experiences:

> I'm single. It's totally a choice on my part, I don't necessarily see myself in a relationship right now, since I happen to move a lot. The farther I am from home, the more I enjoy it, I get to discover new horizons, new people and more things in my work.

Emmanuelle, a 33-year-old single woman from Les Sables d'Olonne, has a lower social status than her parents. Her mother, who is retired, managed social institutions, and her father was a draughtsman and model maker. Her parents 'let her' make the choice of waitressing in restaurants: 'it's my choice, nobody influenced me', she says. To pay for her tuition fees and have 'some pocket money', she began her seasonal worker career as a student. Having received a vocational high school degree in tertiary sciences and technologies with a major in communication (formerly the management sciences and technologies *baccalauréat* degree), she travelled and trained to become a stewardess, and subsequently turned to the food industry. After experiencing repeated failures in the course of her studies, she has followed an erratic, non-linear trajectory. She has been working summer seasons as a waitress on the coast in Sables d'Olonne for four years and says being a seasonal worker 'isn't always easy' in terms of social and familial relationships. Romantic relationships are also 'tough' if the partner does not work in the same professional environment:

> Yeah, of course it does factor in. I think it's tough to be with someone who's not in the restaurant business. In my family, well, yeah, we don't do family celebrations, we miss birthdays, but that's the way it is. Well… it's a choice, I chose this job. I mean, I knew the constraints it involved, and as far as my personal life is concerned, yeah, it isn't always easy. […] Hmm, yeah, one day, sure, I'd like to have a family, but first I need to find my boyfriend, I'm missing that, but yeah, yeah [laughs]. But currently, no, it is compatible. You need to figure out how to do it but it is compatible. Nothing can't be done.

Emmanuelle is aware of the effects of age, particularly when it comes to the physical toll waitressing takes over the years, and is not planning to remain a seasonal worker her entire life:

> Well, actually it depends on whether I manage to get my own business up and running later; I don't see myself running around like I do now at age 50! So I would think it's time to do something else by then. Then again, isn't 50 a bit late for a career change? I don't know, but hmm, yeah. I can't picture myself doing something else, right now I can't see myself doing a different job. [...] I do think you can get tired of it but it depends. I don't know, I think I know that each place is different and if you go to work at a new place, you have different customers, a different way of working. So I don't really know if you get tired of it, even though deep down it's always the same thing. Yeah, I don't know. It wears you out, for sure, I think it's physically tiring.

Considering the preeminence of the open-ended CDI contract as the established norm in working lives and the actual or perceived challenges inherent in reconciling the seasonal worker status and a stable social and family life, what makes seasonal workers make it through a career over the years? Are tourism jobs still part of a career plans, or merely gateways to more 'promising' jobs? Beyond the expected economic rewards of the working contract, a factor that is evoked almost systematically in interviews and which I will not analyse as such here, how do seasonal workers justify their initial commitment to this work and their desire to pursue such careers for a few more years?

2.3 'Passion' as a Modality of Commitment

Many studies have shown that commitment expressed in the rhetorical terms of passion is a specificity of the touristic sector, among a few others. Whether they are passionate about 'old buildings', surfing or sailing (Guibert & Slimani, 2011; Guillaud, 2018), horse riding (Slimani, 2014), travel or 'adventure' (Réau, 2011), tourism professionals pursue careers that are extensions of their own personal practices, of passions they fell for as children or teenagers. How do these professionals negotiate the place their passion plays in their work? Considering what the latter actually involves behind the scenes, and in some jobs, the 'quasi-magical' negation, in the Weberian sense (Weber, 1991), of their market dimension, can we speak here of a disenchantment of the 'relationship to the world' (Réau & Poupeau, 2007), or even of disillusionment?

Analysing the reasons for which they commit to their job 'out of passion' is a practical way to distinguish between categories of seasonal workers depending on their activity sector. While waiters in bars and restaurants and receptionists mostly report being interested in the tasks they perform, those in sports-related service jobs are more likely to say that they are passionate about their work. All the surfing and sailing instructors and the lifeguards I interviewed claim that they chose to became seasonal workers out of a passion that pre-existed their job, citing the love of their sport and of the ocean. The terms 'hobby', 'freedom', 'holidays' and 'beach' frequently recur in interviews – serving as means to contrast these jobs with more traditional, urban and congested professional worlds (Boltanski, 1976). The interviews also attest to a decompartmentalization and a 'more intense blurring of personal and professional life, raising new questions as to the structuring of social times' (Rech & Paget, 2012). Indeed, there is a thin line between a working time consisting in monitoring surfing and sailing activities and a free time when the professional reverts to the status of ordinary practitioner: 'The instructors have gone from consumers to suppliers of sport services' (Bouhaouala & Chifflet, 2001). Claiming to commit to the work 'out of passion' also ultimately entails tacitly accepting challenging working conditions (atypical hours, an intense pace of work in the summer, etc.).

David, who I quoted earlier, is 'kind of the only misfit in the family', he says: his parents, brother and sister are teachers, but the economic stability of his business legitimizes his life choice in his eyes. He received a bachelor's degree in biology and then switched to surfing, earning a professional certificate, after which he decided to make a living out of his passion in Les Sables d'Olonne alongside his family:

> It's the pleasure of passing on what I love, sharing good times in my favourite environment, which is the ocean. I chose this job out of my passion for surfing, yeah, and then it became… It became… At the time, we founded our business, but initially it really was passion. Then we saw that we could make a living from it so we… We, hmm, we jumped straight into it, it became a necessity. […] In the winter I take advantage of all the free time to get rest, enjoy my family, and also to surf and travel.

Pierre is the 21-year-old son of two restaurateurs. He holds a technical certificate (BTS) in tourism and a bachelor's in tourism. He also earned the national rescue diploma (BNSSA). Like his brother, he has been working summer seasons on the main beach of Les Sables d'Olonne for a few years. For him, the distinction between work and leisure is very thin (this is also the case for 23-year-old Anna, who holds a certificate of professional aptitude (CAP) in cooking and makes dairy products in a farm, and says that 'doing seasonal work is combining work and holidays'). This euphemized distinction results in a form of invisibility of labour, which in Pierre's case stems first from his 'enchanted' perception of the job and second from his working environement (the beach):

> I'm doing this seasonal work mostly out of passion. As far as I'm concerned, even if season, even if we're working, I still want it to be a holiday. As a student the financial end of things is the biggest draw, I mean I know this is going to be the part of the year when I'm able to earn the most, compared to the rest of the year. But really, more than the pay, it's passion that explains my choice: being on the beach and being useful. Plus it's not a real job.

Romain is a 31-year-old surf instructor in the summer and physical therapist in the winter. He was born and raised in Les Sables and still lives there, saying he never 'wanted to go anywhere else'. 'I was doing good in Les Sables and I didn't see myself leaving since I had everything I needed here'. He is married and has one child. His parents manage a bar on the Tanchet beach, in the southern section of Les Sables d'Olonne's Corniche road. Romain was socialized in tourism work, or at least touristic economic activity, at a young age through his family. He began working the summer seasons 'at the age of 14/15 and I stopped when I became a physical therapist when I was around 22'. For Romain's family, the tourism economy is a source of work. Doing seasons in les Sables d'Olonne is natural for him, as well as a good way to 'make some pocket money'. Romain therefore combines seasonal work in the summer (he had opened his own surfing instruction business a year before I interviewed him), which is his passion, and a 'real' job 'the rest of the year' in his own practice. Being a seasonal worker is an additional job that he chooses to do, as he earns sufficiently as a physical therapist to 'meet the family's needs':

> I started working at my parents' family bar on the Tanchet beach. I was a sea rescuer for some years and then I became a surfing instructor. After I graduated high school I did a first degree state certificate for surfing, for surfing instruction, and then, hmm, I did two preparatory programmes to get into a physical therapy school, and the school took three years. I did three years of physical therapy training in the Nord region and since then I've been a physical therapist and occasionally I'll attend trainings to perfect my skills. [...] The money's not a motivation for me, since I've got my job to meet my needs; I do it for pleasure. Obviously, I'm not working for free, make no mistake, you've got to pay for the equipment and all that, the truck, the boards, etc. and you have to make a living, right, but I'm not counting on this seasonal work for a living year-round, you know. I've got my job to meet the family's needs.

To Romain, working seasons means 'breaking the routine'. He sees surfing as a 'passion' and a means to support the diversification of his working activities. He freelances as a surfing instructor, which affords him great flexibility as far as time management is concerned in his seasonal work:

> So it's to get to see something different and to deal with a different kind of public; you know, in physical therapy they're patients and for this they're customers, so people do come because they want to. I mean, it creates this whole dynamic, but mostly it's to entertain myself professionally, plus surfing is my passion, and that's worth noting. It allows me to combine fun and pleasure as well.

The modalities of professional commitment that highlight passion are not the only ones to characterize the (more or less lasting) careers of seasonal workers. Being a seasonal worker is also potentially accumulating social relations.

2.4   An Increased Social Capital and a Symbolic Economy

According to some of our interviewees, being a seasonal worker means reuniting with a circle of acquaintances, year after year, that allows to maintain or increase one's social capital. These local networks help them find seasonal work, or at least facilitate access to these jobs. Parties, nights out among seasonal workers are part of this professional world, where friendly and even romantic forms of sociability extend beyond the professional framework. Involving more than just a capital of autochthony, these social interactions also reinforce the sense of 'belonging' as an 'us' (local or visiting seasonal workers, local shopkeepers, etc.), creating a temporary tight knit group for the season, faced with the 'them group' formed by tourists.

Several of the seasonal workers I interviewed reported that they had met their current partner during the season. The porosity of working time and personal time is often singled out as a specificity of seasonal work, especially among the younger individuals. Emmanuelle, for instance, says: 'Now I've built a circle of friends that are also in my line of work. When you're in this line of work, after a while you make friends in this line of work'. Charly, a 26-year-old restaurant worker, says that the working hours that are typical in his occupation impact his social life: 'Even when you go out, you go out with people from the restaurant business because they understand and they live at the same pace'. Pierre, the lifeguard and rescuer at the Les Sables d'Olonne beach cited above, adds that 'when there are parties, the rescuers are often there!'. Thomas, a 26-year-old holder of a certificate of professional aptitude (CAP) and cook in a restaurant in Les Sables d'Olonne, argues that the network built during the season can then be put to use to find work outside of the season, in the mountains for instance:

> 'It's a way to meet interesting people, who know other people, and to develop a large network. If you want to go work the next season in the mountains, you know people who can send you up there, be they colleagues, bosses, people you meet in a bar; it might seem silly but you do meet quite a few people who will allow you to move on to other things in the future.

Advancing age is a variable that is systematically mentioned as a justification of the phasing out of systematic after-work sociability between seasonal workers. François, an ex-seasonal worker who now manages his own surfing school puts it this way:

> Oh yeah, sure, we used to party every night! That was the good thing with the UCPA, this family spirit, where you meet the same people, and the seasons are longer since they manage to offer products from April to late October, and so you can work longer seasons. But that's in the past [laughs]. Now, with the kids, it's not really doable anymore.

Beyond the fact that friends are easily made among colleagues during the season, some jobs in particular (waiter, sports instructor, lifeguard) yield symbolic rewards that allow those who do them to more easily accept long and atypical hours or low pays – in a form of balancing out losses and profits. Service sector jobs that involve dealing with tourists are particularly conducive to such positive perceptions of work, bringing satisfaction, recognition or even a sense of prestige.

Pierre, the lifeguard, is aware of the symbolic rewards of his job, whose representations are in his view characterized by 'respect' and a 'positive image'.

> There's a positive image because we're here for the others. Okay, the point is not to give our lives for someone else, either, that's not the reason, you don't work to die for somebody else, but this helping people out thing is something that people will generally respect.

What 33-year-old Emmanuelle, who I quoted earlier, enjoys in restaurant work is the 'relational side of it, making people happy, and also the fact that there's always something to do, it's an active job actually. You're not sitting at the computer in your office. Yet, she is aware of the 'image' of waitressing jobs, sometimes perceived as 'second-tier'. Despite the low wages paid in the sector, she waves these preconceptions aside, and stresses that working in a restaurant she 'enjoys' is important to her:

I think it's a little bit better than it used to be. I think it really used to be a job where people went like 'okay, you don't know what to do, then you'll be a waitress'; then again, I think some people still do think that. You know, I see the way some customers talk to you, they actually think it's a second-tier job. But I think we… well I don't know, I don't, but I get the feeling that it's better perceived than it used to be. Then again, it also depends in what type of restaurant you work, too. Depending on whether you're just a waitress or an assistant or a manager, or if you're in a snack bar. Clearly, being in a snack bar is not the same as being in a restaurant.

For 21-year-old Chloé:

It depends but people in general do see it as… when they ask 'what do you do?' and I go I'm a waitress, they go 'wow, good for you, you must be tough!'. They tell me I'm tough, that means it's probably not that easy, right!

Julie, the 18-year-old activity leader, finds genuine satisfaction in 'making people happy':

What I liked in being an activity leader was to make people happy, to make them have fun. That's always been the draw for me, I'd say it mainly is, making people happy in various ways. Bringing different people together for shared activities.

Lastly, Quentin, the receptionist, admits he finds it valuable to be 'known' and enjoys the arrangements this makes possible:

Yeah, actually it is always nice to be known and to know people too. Plus, in the small resorts, the small villages, they have a lot of these small seasonal discounts. Everyone knows you, everyone'll buy you something or give you a discount on something. […] I don't know, it was easy to be a seasonal worker in the mountains. There's only young people, people come to party, you meet more people, ten times more.

There is also sometimes an enchanted view of interactions between customers and workers at work here, especially for water sports instructors, lifeguards and sales-workers in some types of shops. The negation of the economic dimension and high-lighting of special forms of sociability and relationships to the body put a sense of prestige into play. However, beyond this 'interest in selflessness' (Bourdieu, 1994), this can also be in the workers' best interest: some customers of sailing, surfing schools and surf gear shops explicitly value this type of social relations.

This is reflected in the words of 21-year-old Steve, a surfer 'since age seven'. He works as a salesman in a surf shop managed by his father. Using the familiar *tu* form [as opposed to the more formal *vous*, which is the rule in such interactions] when dealing with customers comes 'naturally' to him because 'that's the way [he is] in life':

Yeah, we do say *tu* to customers easily. We try to be chill with them because that's the way we are in life and it makes for a more laidback, in-depth conversation with the customer. We can talk more freely. As far as I'm concerned, I use it with all customers, regardless of how old they are.

Jérémy, a 23-year-old high school graduate who also works in a surf shop in Les Sables d'Olonne, is 'passionate about surf' ('I wanted to turn my passion into a job'). He also speaks of the 'state of mind' required for dealing with customers:

> There's a family atmosphere in the shop. The father and the daughter work together. It's super laidback. The customers are local so friendships develop quickly and naturally. And let me tell you, the first thing I was told when I came to work in a surf shop was 'we say *tu* here, in surfing you say *tu*'. We do say that to the new guys, but they would have done it with customers naturally anyway. If they ever worked in a surf shop before, they're used to it. And if on top of that they're surfers, same thing, no need to mention that it works like that to them. We say tu to everyone except the elderly. It would be hard not to say *tu* to young people.

> There's this 'chill' vibe in surf shops, it's kind of like this image everyone has of surfers. And you know what, surfers *are* chill. The most common things you'll hear in a surf shop when a salesman recognizes a surfer coming in are 'Hey, d'you surf today?' or 'How were the waves out there?'

The bodily *hexis* expected of seasonal workers in the field of water sports, which is likely to some extent crafted and controlled, is part and parcel of this symbolic economy. The spontaneous use of the *tu* form, the 'chill' behaviour, the laidback summer vibes, as well as a number of specific physical attributes (tanned skin, muscular bodies, fashionable surfwear, etc.) all contribute to shaping this enchanted commercial relationship: 'By fostering a convivial atmosphere specific to the club or the sport community and highlighting the rhetoric of sports, managers seek to create a sense of belonging among customers and employees, so that they forget the economic nature of their enterprise' (Gasparini & Pichot, 2007). A similar discursive apparatus can be observed in the guesthouses and holiday rentals analyzed by Christophe Giraud, where the tourist must be welcomed as a 'friend', using tokens of familiarity to euphemize the commercial nature of the arrangement: 'Displaying a domestic world is necessary to produce the impression of a convivial, familial and friendly welcome; it allows city dwellers to have a purportedly authentic experience with the region's local residents, an experience in a real family.' (Giraud, 2007). The principles of vision (habitus) of the surf shop salesmen in surf shops and water sports instructors I interviewed, which are adjusted to the world (the field) of surfing, lead them to see their behaviours as evidently natural (*illusio*). While this may be the case, the fact remains that the adopting 'chill' bodily postures and language can be explicitly connected to commercial interests. Some categories of jobs make the 'expression of the self' easier for some seasonal workers (Méda, 2018). As Fabrice, a 44-year old divorced father of two children and ex-seasonal worker who now manages his own surfing school, admits: 'yeah, the tanned, muscular instructor with his straw hat and sunglasses, a friendly and chill guy, it is something that some of the customers are looking for: the myth sells!'.

2.5 Conclusion

This chapter has been an attempt to provide an overview – in other words, a space of possibles – of the reasons for which individuals 'work the seasons' in seaside resorts. While all claim to work in the summer for economic profit, their reasons are not expressed solely in instrumental and utilitarian terms. They diverge when it comes to the idea of 'making it' through the seasons to the extent of doing this work on a long-term basis. Some commit to the work out of 'passion' (like many surfing and sailing instructors); others do it out of necessity (dishwashers and maids). Yet, other reasons for doing multiple seasons also come into play. The friends one encounters every summer are a justification: some workers think of their professional circle as a tight-knit community. In service industry jobs, particularly in bars, servers play the card of social capital (with the customers) and of the benefits of working unusual hours in places they consider 'nice' and 'fun'. Still, with advancing age and the desire for family stabilization (having a stable partner, children), as well as the social expectations surrounding the open-ended CDI working contract, some workers report being eager to stop alternating between seasons of work and unemployment. The sense of 'self-actualization' fuelled by the various benefits expected from seasonal work is rarely permanent, and generally sporadic and discontinuous.

Ultimately, the wide variety of situations it encompasses shows that the term 'seasonal worker' has little analytical value beyond the legal dimension, as it relates to highly variable working conditions in various jobs as well as individuals with highly diverse social backgrounds. Expected (economic, social, symbolic) benefits are envisioned in different ways by the interviewees depending on their social position and sociological background, which partly explain why some will be back working the next season while others will not.

References

Boltanski, L. (1976). L'encombrement et la maîtrise des biens "sans maîtres". *Actes de la Recherche en Sciences Sociales, 2*(1), 102–109.

Bouhaouala, M., & Chifflet, P. (2001). Logique d'action des moniteurs des sports de nature : entre passion et profession. *Staps, 56*(3), 61–74.

Bourdieu, P. (1994). *Un acte désintéressé est-il possible?* (pp. 147–171). Raisons pratiques.

Bozon, M. (2009). Comment le travail empiète et la famille déborde : différences sociales dans l'arrangement des sexes. In A. Pailhé (Ed.), *Entre famille et travail. Des arrangements de couple aux pratiques des employeurs* (pp. 29–54). La Découverte.

Castel, R. (1995). *Les métamorphoses de la question sociale.* Folio Essais.

Castel, R. (2003). *L'insécurité sociale. Qu'est-ce qu'être protégé.* Seuil.

Dethyre, R. (2007). *Avec les saisonniers. Une expérience de transformation du travail dans le tourisme social.* La Dispute.

Ferreira, C. (2006). La 'crainte de la chute': Le retour des classes moyennes dans l'analyse du politique. *Mouvements, 3*(45–46), 168–174.

Gasparini, W., & Pichot, L. (2007). Régulation de travail et culture sportive. L'exemple des entreprises de la distribution d'articles de sport. *L'Homme et la Société, 35–57.*

Gentil, A. (2013). Entre ancrages temporaires et mobilités saisonnières: l'installation permanente des travailleurs mobiles du tourisme sur le littoral atlantique. *Espace Populations Sociétés, 2013*(1-2), 111–124.

Giraud, C. (2007). Recevoir le touriste en ami. La mise en scène de l'accueil marchand en chambre d'hôtes. *Actes de la Recherche en Sciences Sociales, 5,* 14–31.

Grégoire, M., & Join-Lambert, O. (2017). Marges de l'emploi et protection sociale. Une analyse sociohistorique. *Travail et Emploi, 149,* 5–16.

Guibert, C. (2012). Les effets de la saisonnalité touristique sur l'emploi des moniteurs de sports nautiques dans le département des Landes. *Norois. Environnement, Aménagement, Société, 223*, 77–92.

Guibert, C., & Slimani, H. (2011). *Emplois sportifs et saisonnalités. L'économie des activités nautiques : enjeux de cohésion sociale.* L'Harmattan.

Guillaud, É. (2017). Faire face au contretemps pour faire son temps. *Temporalités, 25.*

Guillaud, É. (2018). De l'attrait à l'usure: Les trajectoires professionnelles des éducateurs sportifs en nautisme.

Hoibian, O. (2014). Les professionnels du tourisme sportif de montagne sont-ils préservés du risque de « burn out » ? *Juristourisme, 163*, 24–29. https://doi.org/10.4000/temporalites.3685

Méda, D. (2018). *Le travail.* PUF.

Périvier, H., & Silvera, R. (2010). Maudite conciliation. *Travail, Genre et Sociétés, 24*, 25–27.

Pinna, G. (2013). Vendre du luxe au rabais: Une étude de cas dans l'hôtellerie haut de gamme à Paris. *Travail et Emploi, 136*, 21–34.

Portela, M., & Signoretto, C. (2017). Qualité de l'emploi et aspirations professionnelles : quels liens avec la mobilité volontaire des jeunes salariés en CDI ? *Revue Economique, 68*(2), 249–279.

Première, I. N. S. E. E. (2015). Un million d'emplois liés à la présence de touristes. *Insee Première, 1555.*

Réau, B. (2006). Les devoirs de vacances : La vie quotidienne d'un Gentil Organisateur de Club Méditerranée. *Regards Sociologiques, 32*, 73–81.

Réau, B. (2011). *Les Français et les vacances* (p. 235). CNRS Editions.

Réau, B., & Poupeau, F. (2007). L'enchantement du monde touristique. *Actes de la Recherche en Sciences Sociales, 5*, 4–13.

Rech, Y., & Paget, E. (2012). Les temporalités du travail touristique. *Socio-logos, 7.* https://doi.org/10.4000/socio-logos.2674

Sarfati, F., & Vivés, C. (2018). De l'intérim au CDI intérimaire. Se stabiliser dans le salariat pour limiter la subordination. *Sociétés Contemporaines, 2*, 119–141.

Sebileau, A. (2014). Rester après la saison : l'économie symbolique du néoruralisme balnéaire. *Juristourisme, 163*, 31–34.

Seuret, F. (2007). Les saisonniers, des salariés au rabais. *Alternatives Économiques, 264*(12), 46–46.

Slimani, H. (2014). L'économie de la passion. Formation professionnelle et *turn-over* des moniteurs(trices) équestres sous conditions sociales et affectives. *Actes de la Recherche en Sciences Sociales, 5*, 20–41.

Trancart, D., Georges, N., & Méda, D. (2009). Horaires de travail des couples, satisfaction et conciliation entre vie professionnelle et vie familiale. In A. Pailhé (Ed.), *Entre famille et travail. Des arrangements de couple aux pratiques des employeurs* (pp. 55–78). La Découverte.

Weber, M. (1991). *L'éthique protestante et l'esprit du capitalisme* (p. 285). Pocquet/Agora. (Original work published 1920).

Christophe Guibert is a sociologist, professor at the University of Angers (ESTHUA, Faculty of Tourism, Culture and Hospitality), and researcher at the "Spaces and societies" laboratory (UMR CNRS 6590). For the past 20 years, he has been examining multiple dimensions attached to tourism practices (public policies, jobs, social and cultural uses, gender, etc.) in France, but also in various foreign countries (China, Taiwan, Morocco, USA, etc.). His work is part of a dispositionalist and multi-methodological sociology. He has managed research contracts and published numerous scientific articles relating to these themes. He has published or been editor of *L'univers du surf et stratégies politiques en Aquitaine* (2006, L'Harmattan), *Tourisme et sciences sociales* (2017, L'Harmattan), *Les "sports de nature": une catégorie de l'action politique en question* (2017, Éditions du Croquant), *Emplois sportifs et saisonnalités* (2011, Logiques sociales, L'Harmattan), and *Les mondes du surf, Transformations historiques, trajectoires sociales, bifurcations technologiques* (2020, MSHA). Since 2016, he has managed two licenses and a master's degree in the field of coastal tourism in Les Sables d'Olonne, a delocalized branch of the University of Angers (France).

Chapter 3
Labor, Training and Careers in Tourism in Sardinia

Gabriele Pinna

Abstract In Sardinia, an Italian island region characterized by seasonal seaside tourism, wage labor in the tourism sector is very unstable, not only because of the temporality of tourist flows but also because of the methods of workforce management: whether in small family businesses, managed by entrepreneurs with low cultural capital whose approach to professional relationships tends to be very paternalistic and traditionalist, or in national and international hotel chains, where according to our data, breaches of the law and of employment contracts are systematic, hierarchical relations are quite strained, and youth and women are expected to overwork. Because of these arduous working conditions, students in vocational hospitality schools often decide against pursuing a career in the sector, which contributes to the weakening of the links between Vocational Training and Education (VET) and labor. As a result, most wage workers turn to tourism in a way that reflects an instrumental work attitude; their professional trajectories are very fragmented and often lead, especially for women, to a disqualifying professional integration.

Keywords Sardinia · Working conditions · Job satisfaction · Seasonal tourism · Vocational Education and Training (VET)

This chapter analyses attitudes towards work among wage laborers in the tourism sector in Sardinia (Italy) according to their working and employment conditions. It begins by examining how these attitudes are shaped during the internships of students in vocational hospitality schools. Then, having shown that these internships regularly lead to exits from the sector, it considers the relationships to work of wage laborers that were not specifically trained in tourism, who make up the bulk of the seasonal tourism workforce in Sardinia.

Based on the observation that work constitutes one of the main channels of social integration and identity for individuals, and drawing on a typology developed by

G. Pinna (✉)
University of Cagliari, Sardinia, Italy

© The Author(s), under exclusive license to Springer Nature Switzerland AG 2023
C. Guibert, B. Réau (eds.), *Employment and Tourism*, SpringerBriefs in Sociology, https://doi.org/10.1007/978-3-031-31659-3_3

Serge Paugam (2000), I posit the hypothesis that professional integration in Sardinian tourism fluctuates between the uncertain and the disqualifying form. Reflecting unstable employment working conditions (fixed-term contracts, internships, apprenticeship contracts, subcontracting), relationships to work are defined along a continuum ranging from an interest in the tasks being accomplished, considered by laborers as a source of satisfaction as they allow them to validate and consolidate skills that are central to their professional identities (uncertain professional integration), to a disinterest and in a number of cases dissatisfaction and suffering at work as part of an instrumental relationship to work (disqualifying professional integration).

Job satisfaction relates to job instability but also depends on a wide array of factors. In some circumstances, the job also allows workers to expand on their personal interests and leisure activities, as in the case of sport instructors in France (Guibert & Slimani, 2011) and tour guides in Sardinia. The place of work also matters: in dynamic touristic localities, young workers can enjoy the nightlife during the summer season with their colleagues. The company may or may not provide the organizational resources needed to offer quality services and in doing so prevent its workers from having to deal with customer dissatisfaction (Pinna, 2013). Lastly, age and sex also play a role: job satisfaction decreases as workers get older, particularly among women, who may face mounting challenges in balancing work and family life (Monchatre, 2010).

These features of attitudes towards labor highlight the gap between the respective properties of consumption, a source of distinction and a fundamental part of the tourist experience (Bourdieu, 1979), and of employment, which is often unappealing and demeaning for workers. Overall, employment and labor in the tourism sector are characterized by a gulf between the competencies required to serve a particularly demanding customer base, especially in the luxury sector, and the wages and employment contracts offered for these jobs (Gadrey, 2002). There are indeed many diverse indispensable skills to work in the tourism sector—interpersonal, linguistic (command of one or several foreign languages is expected), in IT; in general, workers must be flexible and multi-skilled, able to balance their work and private lives despite staggered shifts (often at night and on public holidays). Still, despite all these demands, wages in the tourism sector are lower than average (Guégnard & Mériot, 2009), employment is precarious (short-term contracts, apprenticeships, internships, subcontracting), and workforce turnover is particularly high.

Here labor precariousness denotes not only the seasonal and temporary nature of the contracts on offer, but also systematic non-compliance with labor law. In Sardinia, as in the South of France during the summer season (Monchatre, 2006), tourism employees work far longer hours than provided for in their contracts, and overtime is unpaid. Employers are mostly looking to save on labor costs, as a result of which teams are understaffed, and recourse to inexperienced salaried students (Pinna, 2013) and contracting (Puech, 2004) is widespread.

There is also a frequent discrepancy between diplomas and jobs: in effect, a vast majority of workers have no diploma by the end of their training in tourism, and this is exacerbated by the fact that most workers exit the tourism sector after a few years of labor (Molinari, 2010). Unions are seldom present, except in a few hotels that hire a large number of personnel (Beaumont, 2018). Generally, in tourism, workers are evaluated in an individualized manner, and based on their practical skills rather than qualifications obtained through training courses, professional experience and in other professional branches where collective negotiation is better recognized.

The difficult employment and working conditions and lack of recognition of tourism workers are also related to their position within a very rigid gendered, racial and generational social division of labor (Monchatre, 2010; Sherman, 2007). The most prestigious jobs are the preserve of a male workforce whose competencies are better recognized and whose careers are more stable. They are for instance employed as concierges in luxury hotels (Menoux, 2015) and chefs in restaurants. They do not necessarily have academic qualifications, having developed their professionalism on the job, often starting from the lowest rungs of the hierarchy. Still, they enjoy a degree of symbolic capital, and in the case of concierges, of social capital, and are paid higher wages than other workers in the tourism sector. Outside of these skilled work niches, most workers' trajectories, especially those of young people and women, are particularly fractured. Women frequently exit the sector after years of work once they have reached the age of 35 to 40 as a result of their impossibility to balance work and married life, and to a greater extent motherhood. The high turn-over rate of the female personnel and the exits of women over thirty are also the results of choices on the part of recruiters, who prefer to hire young women based on appearance.

A number of studies have also highlighted the impact of a racial division of labor on the difficulty of working conditions and the precarity of work. The tasks that are both toughest and least valued by businesses and customers, such as dishwashing in restaurants and cleaning in hotels, are entrusted to a racialized workforce, made up of Black female immigrants that work jobs such as hotel maid (Ferreira de Macedo, 2003). These workers do the "dirty work" (Hughes, 1996) because the dirtiest and toughest tasks are expected to be performed by those in a subaltern position in the moral order of the hotel and of society as a whole.

Noticeably, in Sardinia, unlike in France and the United States, some "dirty" jobs are still partially the preserve of a local female workforce. These local working-class mothers work in the summers to supplement the family income even though in the more touristic localities we have collected data that show the recourse to foreign female workers to be increasingly widespread. The international sociological litera-ture demonstrates how the most demanding, lonely and urgent, dirty tasks come with a lack of recognition for the workers who perform them (Testenoire, 2010), especially as they are often hired by subcontractors. Many studies have shown that these working and employment conditions contribute to a deterioration of these employees' health (Lada, 2009).

Regarding the subject at hand, I decided to study work and employment in Sardinian tourism after having conducted an ethnographic study in Paris hotels, in which I found that the expansion of the luxury tourism sector is accompanied nei-ther by an improvement in employment and working conditions nor a stabilization of workers' professional trajectories (Pinna, 2013; Pinna, 2018). Drawing on research in summer seasonal tourism work in island regions (Adler & Adler, 2004), I chose to investigate a tourism market that is markedly different from Paris.

Sardinia has a population of around 1.6 million, with a negative demographic trend due to low birth rates and a negative net migration rate, particularly in rural areas, from which many young people, often holding academic degrees, emigrate. This phenomenon is not sufficiently offset by immigration. Sardinia has barely 55.000 foreign residents, which amounts to 3.3 per cent of the population, even though many foreigners come to the island in the summer to work in tourism on a temporary basis. The unemployment rate is higher than the national average and the employment rate is lower, especially for women and young people[1] (even though Sardinia performs better on the main socio-economic indicators than other regions of the Italian *Meridione*).

The Sardinian GDP corresponds to 69% of the European GDP. Speaking in absolute numbers, Sardinia's GDP puts it in the 214th position out of the 281 European regions. Additionally, Sardinia is one of the Italian regions where the most students drop out before completing high school.[2] Although their numbers are low (23.6% in the 25–34 age group compared to 39.9% in Europe), higher education graduates struggle to enter the labor market; they end up doing so at a late stage and often through precarious work arrangements.

In a region where the industrial and agricultural sectors are in crisis, tourism, which amounts to roughly 7 per cent of Sardinia's GDP, is one of the main job providers. Touristic employment, however, is essentially offered during the summer season (82% of arrivals in Sardinia are recorded in the months of June, July, August and September). Over the past twenty years, the number of tourists has increased in particular due to an influx of foreigners (by order of importance, mostly German, French, Swiss and Spanish) fueled by the rise of low-cost airlines. According to employers' groups, there are around 100.000 wage workers in Sardinian tourism, and 80 per cent only work during the four summer months.

The data used for this chapter were primarily collected in the context of a research on the implementation of internships in all Italian upper secondary schools. I conducted ethnographic interviews with principals, teachers, students, and the managers of businesses where they apprenticed (N = 40). Some interviewees were conducted in professional institutes of hospitality. Secondly, I drew on an ongoing comparative research on labor, employment and careers in seasonal tourism in Sardinia and Corsica. I have so far interviewed wage workers in the hospitality and restaurant industries, entrepreneurs as well as tour guides and sports instructors in Sardinia (N = 20).

---

[1] As of 2018, the female employment rate in Sardinia was 42.1 per cent; as of 2017, the unemployment rate among youth aged 15–24 was 46.8% (source: ISTAT, national Italian).

[2] The percentage of NEETs (Not in Education, Employment nor Training) in Sardinia in the 15–24 group is 24.1%, far above the European (10.9%) and Italian rates (20.1%) (Source: 26° Rapporto CRENOS sull'Economia della Sardegna).

The chapter is divided into two sections: in the first, I analyse the professional socialization of students in vocational hospitality schools during their internships. This will evidence how the professional culture and working conditions in tourism businesses ends up discouraging students, who often decide not to pursue a career in the sector. In the second section, I focus on the workers who do not hold diplomas in tourism, in order to highlight their work attitudes and professional trajectories based on the provisional data I have collected so far.

3.1 Internships for Students at Vocational Hospitality Schools

In Sardinia, there are fifteen vocational schools in the field of "Services for enogastronomy and hotel hospitality". As in France, the field is marginalized and stigmatized within the education system, and draws problem students. Due to their low performance in school and lack of interest in theoretical, generalist disciplines, these students turn and are directed by their teachers to programs that are considered less challenging and expected to enable them to enter the labor market quickly (Palheta, 2012). However, vocational secondary education in Italy is strained in many respects: according to the principals and teachers interviewed, its lack of resources and successive reforms have turned vocational schools into third-level high schools. Less time is now devoted to practical teachings and lab work, whereas more is spent on the humanities and math, a trend that does not appear to fit the needs of students.

Another point often comes up: unlike in France, whose case was spontaneously cited by interviewees, Italian vocational hospitality schools do not have the means required to open restaurants or even hotels for teaching purposes, which would enable students to experience customer service work without exiting the institutional educational framework. In Sardinia, among the fifteen institutes, only one has a restaurant for training. Additionally, the national vocational stream increasingly faces competition from regional vocational programs that grant three-year secondary level degrees.

In 2015, Law 107 (the so-called "Good School" act) introduced another reform of the national educational system, providing for the introduction of internships in all secondary schools (high schools as well as technical and vocational schools). The underlying principle of the reform was to ensure that training in the workplace has the same value as the training offered in school classrooms and labs. Job training was also expected to improve career guidance and chances' students of entering the labor market. Unlike in the mainstream high schools (devoted to the humanities and the sciences), in technical and vocational schools the reform was welcomed by most teachers and principals, and these institutes received increased resources to strengthen their ties with businesses in their respective sectors (Pinna & Pitzalis, 2020).

In this section, the development of the connection between vocational schools and the tourism sector is highlighted. Paradoxically, while students in hospitality schools are quickly socialized at work, it happens quite frequently that they decide not to pursue careers in tourism. Indeed, job experiences during internships are often painful confrontations with the reality of the labor market. This does not lead, however, to "conversions", such as those that have been described by interactionists in studies on healthcare work, for instance (Davis, 1968), but rather to exits from the tourism sector, or at least from tourism in Sardinia.

Teachers, as I have noted, support early entry professional integration for these students. Considering most students come for working-class families (which often struggle considerably as a result of the father's unemployment), teachers believe that entering the labor market can in itself be a form of upward social mobility:

> That's the first thing we see here, we have students from very poor, underprivileged social classes, so when they get here, I remember the first restaurant class I taught with a colleague five years ago, they didn't even know fish forks existed, they didn't know fish either, like seabream, they'd only had cod. There was a whole bunch of things, for instance when we had they wear the uniform, they suddenly understood what it is to play a role, to do something with their lives; we removed them from a family situation where they were stuck, where everything is always the same, they struggle to make it to the end of the month.

Teachers' discourses on their students sometimes reflect class prejudices, leading them to assume that their students' social backgrounds and low grades justify their professional outcomes, meaning that they should be glad to perform menial jobs with little social value. Teachers are aware of the difficult working conditions students face in the course of their apprenticeships as well as during the summer seasons, which are often an extension of these first student jobs:

> They realize it's a very tough job; when they work the season, they're paid eight hours but actually they're working 12-14 hours a day, it's punishing work. (A teacher at a vocational hospitality school)
>
> The discipline was very strict, the chef was really hard on them, he's pretty much a dictator, a king, there's no democracy in the kitchen, in an upscale hotel. I went with a colleague to accompany the students, and for me it was.... we went to dinner, I'd never seen such discipline in a dining hall, the distance between the maître d' and the busboys is huge. Students just couldn't believe it at first. (A teacher at a vocational hospitality school)

Such discourses also appear to justify and in a sense legitimate not only the toughness of the students' working conditions but also the gender inequalities that are reflected by a very rigid gender division of labor and a macho culture. Young women discover a professional environment in which they are relegated to the lowest-valued jobs (such as waitress) and must learn to cope with sexist remarks, sexual propositions and in some cases harassment by colleagues and supervisors (Pinna, 2015). Regarding a student who reported to her parents that she was harassed during her training, a (female) teacher told me:

> It's ugly, sometimes, what these girls [...] it made me very sad (*the student's behaviors*); employees told her things, adults, yes, because you see there's a locker room where they

change, but these things happen in that environment, it's not fair to call it harassment… (A teacher at a vocational hospitality school)

Managers expect students to be able to work autonomously after mere days of training of the job, even when they are very young (16 to 18).

Teachers are thus well aware that businesses cannot offer high-level training to students. This is why they attempt to identify managers with an awareness of the importance of quality training to send them their best students.

This turns out to be a difficult task. In the Sardinian tourism sector, businesses are often very small and family-run, and hire a small number of personnel. Bosses have little cultural capital and have neither the resources nor the professional culture needed to arrange for high-level training. They mostly appear to be looking for a free workforce to augment their staff:

> Many businesses volunteer for internships, and then you can clearly see whether the company is serious about it. You have companies that are just looking for manpower; sure, it's give and take, you can't deny that students in kitchens and halls are giving a hand, but at the same time the manager has to offer them training and supervise them, they can't just give them the same posting for three weeks straight, that's not how it's done. (A teacher at a vocational hospitality school)

In large hotels and upscale tourist resorts, managers pursue the goal of reducing labor costs to boost profits. Recourse to student labor is very frequent, but high-level training is not necessarily offered. These recruits are often entrusted with the "dirty work" (Hughes, 1996):

> We've sent very gifted student cooks with very good attitudes towards work (*to a very renowned luxury hotel*), they (*the managers*) asked them to perform very simple tasks, menial labor, right… the students wanted to see and do more, because they're still in school, they're not kitchen helpers, and they asked them to clean mullets for days on end, and the students ended up complaining. (A teacher in a professional institute of hospitality)

The massive recourse to students also destabilizes the tourism labor market by lowering the number of opportunities available to professionals:

> You have businesses that hire entire classes to work the season; some principals turned them down, it was what we'd call "summer internships", students would be compensated at a level of 300-400 euros and they were hired for two or three months, it did some harm, many principals wouldn't send them because under those arrangements there was exploitation, and also very few professionals who wanted to work were hired, teams were understaffed because of the intern students" (A teacher in a professional institute of hospitality)

Experiences in the field are thus very often negative for students, who are confronted directly to difficult working conditions, strict and paternalistic hierarchic relations and a professional culture that values the exploitation of young employees, be it in small, family-run businesses or in hotels and tourist resorts managed by national or international choices. Logically, many of them decide not to pursue their careers in the sector further once they have obtained their vocational certificate in hospitality.

3.2 Work Attitudes and Careers in Tourism

On the one hand, the internship phase of training and the experience of work in the tourism sector in general may discourage students in the vocational hospitality stream from continuing in the field, thereby contributing to the weakening of the school/work connection; on the other hand, my ongoing empirical research has highlighted that numerous students in vocational institutes (in fields other than hospitality) and at university frequently work in tourism-oriented food and hospitality businesses. These are students from working-class backgrounds, with families possessing little economic capital. They work as waitstaff, assistant cooks, receptionists, night watchmen, but also as lifeguards and sport instructors during the tourist season. In most cases, they turn to the tourism sector for instrumental purposes, to pay for their studies or acquire greater autonomy from their families. During interviews, these young students display a very uncommitted work attitude: they do not spontaneously discuss the tasks they perform, claim that all student jobs are hard and that what matters is getting paid. Their careers in tourism are very precarious, and even if they work regularly during the summers, they very often switch companies:

> It's not a job to me; I just told myself I needed to make a little extra, I needed to go out, have new experiences, I needed to escape, as well, I'd always lived in my village until I was nineteen. (A student worker)

Difficult working conditions and exploitation are accepted because they are perceived as temporary. These students will also readily cite positive aspects of these jobs, such as the opportunity to interact with wealthy customers (Peretz, 1992), and more generally to work and live in posh, highly renowned touristic locales like the Costa Smeralda. In such settings, student workers can develop friendships with their colleagues and enjoy the summer nightlife, which leads some to view the experience positively, despite their accumulated fatigue from work overload and lack of sleep. Relationships with peers play a decisive role in that they counterbalance the brutality of workplace relationships in hotel and restaurants.[3]

Even though the main motivation of these students is financial, the weight of these other factors, such as the eagerness to experience a more dynamic environment than in the places where they came from, should not be underestimated. For students, working in tourism during the summer season is a challenge, a trial by fire, considering the difficulty of the conditions, and making it to end of the season allows them to strengthen their own identity. It should however be recalled that over-investment in work during the season and the subsequent need for rest have a negative impact on performance at school—a phenomenon that has also been evidenced in France for student workers (Pinto, 2014).

---

[3] A number of interviewees noted that conflicts with managers and bosses contribute to the very high worker turnover rates observed during the season.

Regarding female student workers and young female workers in general, who work mostly as waitresses and receptionists, and sometimes as maids, they are on the wrong end of cumulative inequalities, as numerous sociological studies have shown (Ferreira de Macedo, 2003; Puech, 2004; Monchatre, 2006; Monchatre, 2010). Their over-investment at work is trivialized, and managers assign them numerous tasks, including supervisory ones. Over time it becomes more and more difficult for women to balance the demands of seasonal work with those of marital relationships and motherhood. Young mothers frequently entrust their children to loved ones during the season, particularly when they have to work far from home. Their work attitudes are ambivalent, however: despite obvious struggles, in a region characterized by particularly low employment rates for women, they may perceive this seasonal and precarious work in a positive light in that it gives them economic independence from their partner or family.

To deepen the analysis of the connection between tourism labor and workers' professional trajectories, I decided to interviewed older workers (aged 40–60), who have been working the touristic season for many years, albeit under temporary contractual arrangements. Employment instability is a difficult experience for adults who have families with children. Unemployment benefits are insufficient to ensure decent living conditions for their families year-round, and they have to implement a number of strategies to secure additional resources, such as partaking in informal activities (hunting, fishing, home vegetable gardening), living in a rural area where they can afford a family home, or where at least rents are significantly less expensive than in coastal, tourist towns. Adult parents seldom work both touristic seasons (summers in seaside resorts and winters in ski resorts); this is more frequent among younger, single workers.

The older workers I interviewed are all men and work as chefs in restaurants and night watchmen in hotels. Their relationship to work can be described as negative in that they all express a preference for stable employment, which they say is impossible to find in Sardinia. Yet, it can also be positive, especially among chefs, who highlight their commitment, having worked hard since they were very young and climbed up the ladder since they began as assistant cooks. They describe this investment at work in terms of a vocation, from which their skills and qualities result. In that sense, these jobs are akin to craftsmanship to them.

Yet, recent trends in the tourism labor market have had a very negative impact on their working conditions and fragilized their representations of these jobs. Interviewees are bitter about the degradation resulting from a massive recourse to student labor and foreign workers, who only stay in Sardinia for the season. Among these older workers, an interesting case is that of those who, unlike the chefs, turned to tourism after a career in another professional field, often following a period of unemployment. These men work as receptionists, night watchmen or handlers. They do not experience their work as a vocation, and merely refer to it as the only opportunity they got to reenter the labor market.

3.3 Conclusion

The predominance of an instrumental work attitude and of professional trajectories within a "transitional labor market" (Monchatre, 2010), especially for women and young people, along with the weak connection between the job market and the education system, is a result not only of the structurally seasonal temporality of employment, but also of the very tough working conditions that prevail in the sector. Heavy workloads, systematic breaches of contracts, and the trivialization of over-investment in work create very tense relationships with supervisors.

On the one hand, in small family-run restaurants and hotels, workers have to deal with bosses that are sometimes described as "despots", as they expect total commitment from their employees while resorting to paternalistic management methods and often trample on labor laws and workers' dignity. On the other, in companies that belong to national and international hotel chains, workers report a degradation of working conditions resulting from the obsessional efforts to cut down on labor costs, with widespread recourse to intern students and apprentices, foreign seasonal workers and subcontractors. Despite the expansion of the tourism labor market that has followed the increase in international tourism over the past twenty years, as I have already show regarding Parisian luxury hotels (Pinna, 2013; Pinna, 2015), the numerical increase in jobs does not ensure their quality (Gadrey, 2002). Unsurprisingly, the toughness of their working conditions leads students from professional hospitality institutes to turn to other sectors, as if for them, although they come from underprivileged backgrounds, these tourism jobs were incompatible with school.

Even if mass unemployment and the precarization of Sardinian society still ensure the availability of a workforce for tourism businesses, in terms of the typology evoked in the introduction to this chapter (Paugam, 2000), the succession of precarious professional experiences may lead to a disqualifying integration. Indeed, most workers rely on tourism only for instrumental purposes. While the younger ones are chiefly looking to earn money to fund their studies and their social life, the older workers often use it as their only available source of income and way out of unemployment.

Workers have difficulty strengthening skills that are formally recognized on the labor market and stabilizing their professional trajectories: women, especially, who are employed in the less prestigious positions (maid, waitress), are at risk of exiting the tourism sector to turn to even less skilled jobs in cleaning and home care (Monchatre, 2010).

Tourism, which is constantly brought up by politicians and experts as the main driving force of the Sardinian economy and the cornerstone of its future economic development, neither offers quality jobs nor ensures professional integration (meaning employment stability and job satisfaction, per Paugam's typology). Also, even though compared to other tourism hotel markets, such as the United States (Sherman, 2007) and France (Pinna, 2018), the number of foreign workers is lower, I have observed phenomena that resemble those that have been evidenced in other touristic island regions (Adler & Adler, 2004). Low-skilled, badly paid positions are filled by local workers whereas executive and management positions are reserved to those from continental Italy. Ultimately, there is a dramatic gap between the social prestige associated with touristic consumption, particularly in places like Sardinia, with its pristine beaches and hip luxury spots, and the quality of work and employment for those who help tourists live these dream experiences (Poupeau & Réau, 2007).

References

Adler, P. A., & Adler, P. (2004). *Paradise laborers. Hotel work in the global economy*. Cornell University Press.

Beaumont, A. (2018). Tirer parti de l'ordre établi ? Les socialisations politiques au travail dans un hôtel de luxe. *Politix, 2*, 79–105.

Bourdieu, P. (1979). *La distinction. Critique sociale du jugement*. Éditions de Minuit.

Davis, F. (1968). Professional socialization as subjective experience: The process of doctrinal conversion among student nurses. In H. S. Becker et al. (Eds.), *Institutions and the person: Essays in honour of Everett C. Hughes* (pp. 235–251). Free Press.

Ferreira de Macedo, M.-B. (2003). Femmes de ménage et veilleurs de nuit : une approche sexuée du travail précaire dans un hôtel en France. *Cahiers du Genre, 35*, 189–208.

Gadrey, J. (Ed.). (2002). *Hôtellerie-Restauration : héberger et restaurer l'emploi*. La Documentation française.

Guégnard, C., & Mériot, S.-A. (2009). Alice au pays des hôtels : de l'autre côté du miroir. In E. Caroli & J. Gautié (Eds.), *Bas salaires et qualité de l'emploi : l'exception française ?* (pp. 269–332). Presses de l'ENS.

Guibert, C., & Slimani, H. (2011). *Emplois sportifs et saisonnalités. L'économie des activités nautiques : enjeux de cohésion sociale*. L'Harmattan.

Hughes, E. C. (1996). *Le regard sociologique : essais choisis*. Éditions de l'EHESS.

Lada, E. (2009). Divisions du travail et précarisation de la santé dans le secteur hôtelier en France : de l'action des rapports sociaux de sexe et autres rapports de pouvoir. *Travailler, 22*, 11–26.

Menoux, T. (2015). La distinction au travail : les concierges d'hôtel de luxe. In M. Quijoux (Ed.), *Bourdieu et le travail* (pp. 247–266). Rennes.

Molinari, M. (2010). *L'insertion des jeunes dans l'hôtellerie-restauration*. Net-Doc.

Monchatre, S. (2006). Instrumentalisation des femmes au travail et du travail par les femmes de l'hôtellerie-restauration. In E. Flahault (Ed.), *L'insertion dans tous ses états. Formation, emploi et travail des femmes* (pp. 231–242). PUR.

Monchatre, S. (2010). *Etes-vous qualifié pour servir ?* La Dispute.

Palheta, U. (2012). *La domination scolaire. Sociologie de l'enseignement professionnel et de son public*. PUF.

Paugam, S. (2000). *Le salarié de la précarité*. PUF.

Peretz, H. (1992). Le vendeur, la vendeuse et leur cliente. Ethnographie du prêt-à-porter de luxe. *Revue Française de Sociologie, 33*(1), 49–72.

Pinna, G. (2013). Vendre du luxe au rabais: Une étude de cas dans l'hôtellerie haut de gamme à Paris. *Travail et Emploi, 136*, 21–34.

Pinna, G. (2015). Luxe, genre et émotions dans l'hôtellerie. *La Nouvelle Revue du Travail[Online], 6*. http://nrt.revues.org/2135

Pinna, G. (2018). *Travailler dans l'hôtellerie de luxe. Une enquête ethnographique à Paris*. L'Harmattan.

Pinna, G., & Pitzalis, M. (2020). Tra scuola e lavoro. L'implementazione dell'Alternanza Scuola Lavoro tra diseguaglianze scolastiche e sociali. *Scuola democratica. Learning for Democracy, 11*(1), 17–35.

Pinto, V. (2014). *À l'école du salariat. Les étudiants et leurs « petits boulots »*. PUF.

Poupeau, F., & Réau, B. (2007). L'enchantement du monde touristique. *Actes de la Recherche en Science Sociale, 5*, 4–13.

Puech, I. (2004). Le temps du remue-ménage. Conditions d'emploi et de travail de femmes de chambres. *Sociologie du Travail, 46*(2), 150–167.

Sherman, R. (2007). *Class acts: Service and inequality in luxury hotels*. University of California Press.

Testenoire, A. (2010). Articuler les paradigmes de la reconnaissance et de la redistribution ; la situation des femmes de chambre. *L'Homme et la Société,* (2), 83–99.

Gabriele Pinna is researcher at the University of Cagliari in Sardinia (Italy). Gabriele Pinna received his PhD in sociology at the University of Paris 8. His main research interests are work and employment in tourism and vocational education and training (VET). He has published articles in French, Italian and English in various international scientific journals and a book in French entitled *Travailler dans l'hôtellerie de luxe. Une enquête ethnographique à Paris* (2018).

Chapter 4
Who Benefits from Tourism? The Ambiguities of Development Through Tourism for Water Sports Instructors

Etienne Guillaud

Abstract Tourism is, for nautical sports teachers, the object of speeches that insist on the benefits (potential or proven) both for their conditions of employment and work as well as for the economic and symbolic impact on the territories where they work. Indeed, although sports educators in charge of supervising these activities may have an interest in the development of tourism because it justifies their employment (and their perennity), they are not equal in this process. This promotion of tourism is not always matched by recognition of the supervising work, which is necessary for the production of the tourist offer, and thus contributes to reinforcing inequalities within the professional group.

Keywords Professional group · Seaside tourism · Water sports

4.1 Introduction: Is Tourism a Boon for Water Sports and Employment?

'The French Sailing Schools are here to make you experience unique emotions, in a completely safe environment, and in a fun, congenial atmosphere!', reads the website *Sur l'eau, tout est + fort* [On the water, everything's stronger],[1] a communication campaign that brings together fifteen centres in the Pays de la Loire region of France that have received the "sailing school" label (which guarantees the quality of their supervision of sailing activities). The campaign was launched by the sailing league (the regional branch of the French sailing federation) as part of a training course on web marketing for sports educators. The goal of the course was to have

---

[1] Sur l'eau, tout est + fort [online]: http://www.toutestplusfort.com/ (last accessed 2 July 2020).

E. Guillaud (✉)
CRBC, UBO, Brest, France
e-mail: etienne.guillaud@univ-brest.fr

C. Guibert, B. Réau (eds.), *Employment and Tourism*, SpringerBriefs in Sociology, https://doi.org/10.1007/978-3-031-31659-3_4

43

them master the tools specific to this new component of the job. The website *Sur l'eau tout est + fort!* was an offshoot of this, complete with the #PlusFort (#stronger) hashtag on social media, with logistical and financial support from the region. Such communication campaigns, much like labels and sailing events, serve as 'sounding boards' for water sports (Bernard, 2016). They reflect a shared interest among water sports schools that seek to attract new customers and local political actors who seek to promote their territory. The 'surfing city' label is for instance an opportunity for some seaside resorts to symbolically associate themselves with a practice whose positive image may in turn boost their own (Guibert, 2006) and to promote their beaches, which directly benefits institutions that offer boardsports activities. On a different level, organizing a land sailing championship allows both water sports institutions and local authorities to promote that practice as well as its 'local champions', and thus to attract individuals with distinctive backgrounds (Koebel, 2011) on practice sites. The existence of this shared interest between water sports schools and local policy actors points to the porous distinction between public and private interest (Faure & Suaud, 2009) in spaces (such as coastal areas) that are characterized by an economic, symbolic and temporal dependence on tourism. In addition to a sports federation whose communication is partly a source of market-oriented profits for private actors, one may for instance find on the coast private, for-profit water sports organizations with an interest in offering 'disinterested' services (Bourdieu, 1994) for promotional proposes – say, free lessons during a sailing festival organized by a municipality.

On the coast, at least, water sports are highly dependent on the touristic draw of the area in which the schools are based. Tourism gives water sports instructors a material and symbolic existence. It gives them a material existence in the sense that they derive their income from commercial activities that are directly tourism-related, especially during the summer months. 'Tourism development' is accordingly associated with the creation or stabilization of jobs in water sports centres. It gives them a symbolic existence in that it contributes to legitimizing their place locally, especially for new country dwellers, and in that they perceive themselves – at least those in the most stable situations – as actors of 'tourism development'. Tourism also contributes to making the job attractive on the market. Providing services that focus on the customer's enjoyment, working on the beach in the summer season in the 'fun environment' of a water sports facility makes the job a socially and symbolically rewarding endeavour, at least initially. All of these factors are conducive to an 'enchanted' (Réau & Poupeau, 2007) relationship to work, liable to relegate the 'objective truth' (Bourdieu, 1996) of the job to the background.

Water sports instructors tend to speak of tourism in ways that highlight its (potential or effective) benefits, in terms of employment and working conditions as well as of economic and symbolic rewards for their local communities, even though the seasonality inherent in tourism contributes to job precarity for instructors in the field of recreational physical activities (Guibert & Slimani, 2011). This chapter examines this belief in the benefits of tourism development by drawing on a survey of the professional group of water sports educators conducted as part of a PhD research on the coastline of the Pays de la Loire region between 2013 and 2017. This research was mainly based on direct observations of work in water sports facilities and on 63 semi-directive interviews[2] that mapped out a space of viewpoints on the job and its challenges. What are the effects of the increasing role of tourism in water sports activities on instructors? In the process of addressing this question, I was able to shed light on the ambivalences inherent in this development as well as on the way in which the prevailing orthodoxy on the benefits of tourism masks differences of interest within the professional group.

Located in the Northwest of France, south of Brittany, the coastline of the Pays de la Loire region spreads over 450 kilometres across the departments [sub-regional levels] of Vendée[3] and Loire-Atlantique. The region is openly counting on the coastline for its economic development – especially on resorts such as La Baule, les Sables d'Olonne and on the island of Noirmoutier, which attract many summer vacationers. The map below (Fig. 4.1) combines the status of the centres – private and for profit, private and non-profit, public or semi-public[4] – and the number of activity categories offered within them among the following: sailing, land sailing, surfing, kitesurfing, and canoeing/kayaking. This number is generally correlated with the number of paid positions within the centre.[5] It highlights the existence of two types of configurations in municipalities. In the first – as in Saint-Jean-de-Monts – water sports instructors tend to work in an often large, multi-activity facility with a virtual monopoly in the sector. These are particularly concentrated in the Northern part of the Vendée and in the Southern part of the Loire-Atlantique coastline. In the second, water sports instructors are spread out over multiple, mostly private facilities, offering one or two activities. This is particularly the case in municipalities where private surfing schools proliferate on the beach in the summer season.

Here, the term water sports instructor encompasses people who supervise sailing, land sailing, surfing, kitesurfing, canoeing and kayaking activities,[6] and who

---

[2] This includes 48 interviews with currently or formerly active water sports instructors and 15 interviews with actors working on the periphery of this professional groups (institutional actors, employees in non-instructional positions, etc.).

[3] For more detailed data on tourism in Vendée, see Christophe Guibert's contribution to this book.

[4] The term 'semi-public' applies to facilities who are dependent on a public authority even though some of their funding comes from private sources, operating for instance as semi-public companies (SEM) or industrial and commercial public establishments (EPIC).

[5] The counting method used here does not make it possible to establish the number of paid employees on the coastline, which would require more research, but it gives us an idea of it. Empirical research suggests that the higher the number of activities offered, the higher the number of paid employees. If, for instance, a facility offers activities that can be supervised by the holder of a single diploma, it is likely to employ only one or two workers. On the other hand, if the activities offered by the facility require five different diplomas, it is likely to employ at least four full time-equivalent workers.

[6] These should all be considered as categories of activities: a sailing instructor might teach windsurfing, catamaran and dinghy courses; a surfing instructor might teach surfing and other board sports such as stand up paddleboarding. The 'water sports' category itself, like the 'outdoor sports' category, lends itself to criticism as it 'assembles dissimilar practices' (Audinet, Guibert, Sébileau, 2017). That said, relying on this category in this chapter does not seem inappropriate insofar as

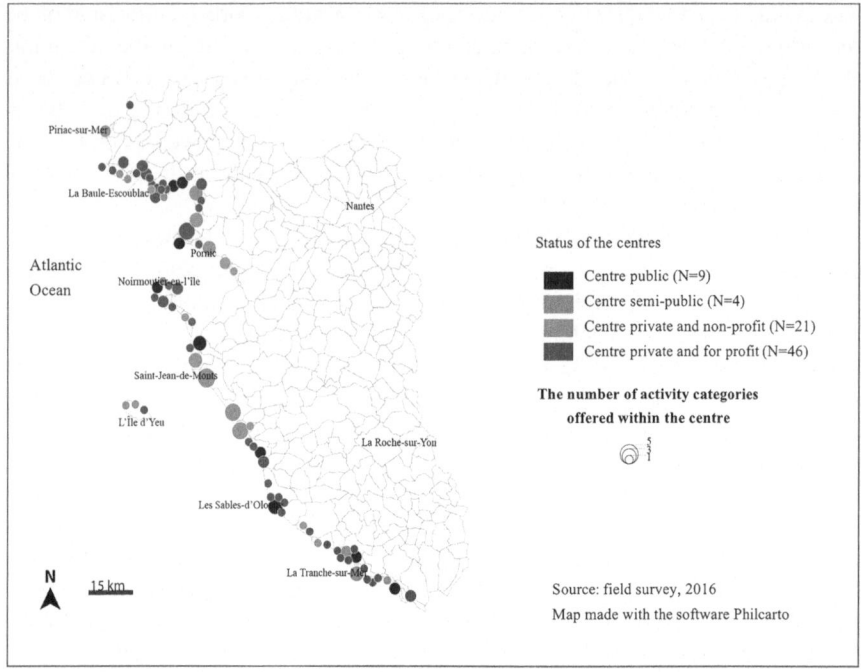

Fig. 4.1 Spatial distribution of nautical centres on the coastline of the Pays de la Loire region

are holders of a certificate allowing them to be paid for this work year-round. Federal instructors are a different category, subjected to limitations on paid employment. On the coastline of the region under study, the number of sports instructor positions increased and stabilized from the mid 1990s to the early 2000s, thanks to employment supports schemes such as the *emplois jeunes* [youth employment contracts]. In addition to making it easier to hire personnel, these schemes led to a redefinition of the professional activity (Demazière & Pélage, 2001), ensuring its integration in the tourism market especially through the 'rationalized' management of centres or the creation of new types of supply. Water sports instructors are mostly men (around 90% according to counts made in the centres under study), with varied social backgrounds, often middle-class (sons of teachers, healthcare professionals, salespersons, craftspeople, etc.). This job was rarely their first choice: generally they found themselves frustrated or facing failure (academic failure, failure to find a job, or experiences of boredom at work or conflict with an employer, etc.), which led them to pursue a different professional career path (Denave, 2009) endowed with symbolic value, especially thanks to the proximity of the sea (Guibert & Sébileau, 2014).

instructors in the aforementioned activities generally have a sense of belonging to the same professional group and as the category is used by institutions (such as municipalities and tourism offices) for communication purposes.

The relationships between touristification and water sports instruction jobs can be analyzed from two angles, reflected in the two parts of this chapter. The first part will analyze the effects of tourism on the work of sports instructors. These include the deepening of divides within the professional group, between the most stable and the most precarious workers. Then, the second part will draw on a case study of the town of Saint-Jean de-Monts to show how local authorities encourage the touristification of water sports while paradoxically remaining fairly indifferent to the working conditions of water sports instructors. The question of tourism will ultimately call into question the process whereby the work of water sports instructors is made visible or invisible (Krinsky & Simonet, 2012), which partly determines whether or not they aspire to stick with the job.

4.2 Heightened Seasonality and Diversification of Activities: The Ambivalent Effects of Tourism on the Professional Group

There is undeniably a form of economic rationality at work in the defence of touristic development as a means of employment development in water sports instruction on the French coastline, particularly during the summer season. The mainly tourist-oriented commercial activities of water sports facilities constitute the bulk of their revenue. Other activities geared towards members of sports associations and socio-educational activities (for students especially) may justify the social role of the instructors and give them work that is not dependent on the temporality of tourism, but they are poorly paid and generally poorly subsidized by local authorities. Subsidies have even tended to decrease for summer camps and *classes de mer* [class trips to the seaside], which may include such activities. Tourism is thus objectively the main resource for creating and supporting jobs. Still, the effects of tourism on the nature of these jobs, on the work itself and on the organization of the professional group have not received much attention.

Tourism has three observable effects on water sports instructors. First, it determines the seasonal variations of their professional activity, which follows an unusual schedule compared to the dominant temporal norms in salaried work. 'We work when everyone else is on holiday', interviewees told me repeatedly. Salaried instructors on an open-ended contract might do 60 per cent of their annual hours between April and September, including 30 per cent in July and August. Summer overtime is compensated by time off in winter. An independent instructor, the owner of his own facility, told me during an interview that he works only between March and October, but with workweeks of up to 6070 hours in July and August. These seasonal variations partly determine the nature of the jobs. With the exception of a few very small private for-profit establishments (especially specialized in surfing and kitesurfing), most water sports facilities – associations or semi-public organizations – on the coast have a few salaried instructors on open-ended contracts and a majority of instructors on short-term contracts (5–8 months). Precarity is thus quite widespread in the profession. The variations in the pace of work over the year are a challenge both physically and in terms of juggling work with family life, a task for which the instructors are unequally endowed.[7]

Tourism also contributes to the diversification of tasks at work. While water sports instruction remains the bulk of the job, the touristic character of this activity also requires dealing with communication and management (in terms of planning, equipment, billing, etc., but also of management of seasonal worker teams). Such diversification opportunities tend to be perceived as rewarding, being means of escaping professional routines that help fend off fatigue.

Lastly, tourism contributes to strengthening a number of professional norms owing to its 'relational' or 'interactional' dimensions (Cartier & Lechien, 2012). The skill of mastering interactions is specific to service work and based largely on informal know-how (Faguer, 2007). It can turn out to be useful to build a customer base among tourists. It can also be symbolically valued on the job market ('the public likes him'). Water sports centres seek to benefit from this skill economically, especially by advertising comments praising the team for being 'friendly', 'smiling' and providing 'quality service' on social media. Yet, the instructors also have to exude rigour so that their services are concurrently perceived as 'professional' and safe. Professionalism also requires distancing oneself from the turn-off image of the instructor 'who 'chills on the beach', 'parties too hard', or is only there to 'hit on girls'. This may involve efforts on the presentation of the self (Goffman, 1973) and strict compliance with regulations. The discourses and practices pertaining to these interactional norms reflect a concern for taking sports instruction jobs seriously, making it possible to work year-round. This in turn reflects an effort to initiate a movement whose goals include ridding the profession 'of people who are not mobile enough to go along with the changes' (Hughes, 1996).

Relationships to these effects of tourism differ according to one's position in the professional space, along a continuum between those in the most stable positions, especially full-time employees on open-ended contracts and the least stable, such as seasonal employees on fixed-term contracts.[8] The latter have the least resources to protect themselves from the effects of seasonal variations in activity. They face more uncertainty for their future, and have reduced options (like adopting flexible hours, taking time off) to reconcile the mismatched temporalities in their lives. Indeed, while the touristification of water sports results in the diversification of tasks at work, the latter are subject to a distribution between teams. Often, those in the most stable positions in the centre tend to hoard the most rewarding tasks, especially those having to do with management and communication, and to leave the

---

[7] I have expanded on this aspect for the journal *Temporalités* (Guillaud, 2017).

[8] Seasonal workers can also occupy different echelons in the professional space depending on whether they are on longer (7–8 months) or shorter contracts (5–6 months), on the number of activities they are allowed to supervise, on whether they are 'return' workers who have come back to work for the same establishment several years in a row or 'newcomers', on the extent of their professional experience, of their athletic achievements, etc.

'dirty work' to others. They find it easier to fend off fatigue at work by diversifying tasks and avoiding the most physically demanding ones (especially those that require handling equipment or regularly dealing with rough weather and sea conditions). They tend to less often deal with the publics considered least rewarding – the beginners who have come to discover a sport while on a holiday. Lastly, seasonal workers are expected to comply with professional norms to a greater extent, since it is in their interest to secure the loyalty of their centre for greater stability (or even a full-time position). They accordingly have to abide by the incentives inherent in tourism-induced commercial relationships ('smile and be friendly', 'be professional', 'be hard-working', etc.); failing that, their local reputations could suffer. Conversely, the most stable workers are in a position to produce the professional judgments that partly determine the employment situation of their least stable colleagues. Tourism may be a key factor in the professional group, but it does not benefit everyone in the same way.

The differences in relationships to the effects of tourism highlight the existence of professional hierarchies, based especially on the divide between full-time and seasonal workers. These hierarchies are sometimes reflected in financial inequalities, as a seasonal sailing and land sailing instructor (on a six-month contract) explained to me. She complained about the unequal distribution of bonuses and days off, to the benefit of the full-time staff, even though seasonal staff sometimes actually work more during the summer season, and are most likely to deal with the less experienced tourists, a task that is considered less rewarding:

> Anyway, it goes like this here: when the full-time staff have to prepare the schools for the summer, we [the seasonal workers on longer contracts] do all the classes. So I do class after class, especially since I've been doing land sailing [...] ever since I've been able to do both, I've been working a lot more, and obviously that's valuable to my employer [...] I can see that the full-time workers all have lunch vouchers, a thirteen month and all that stuff; [the pay] doesn't change, but they have those perks. They get a seasonal bonus that we don't have. [...] Last year I did land sailing, cruising and windsurfing, and I got zero raise. Whereas in effect I was doing three jobs, and I wasn't paid any better. This year, actually, they used a software to make stats about how much the seasonal staff and the full-time staff work on classes. And they've found that the seasonal staff does more than the full-time staff! [...] It doesn't create tensions, but we just get a bit sick of it sometimes. We don't get the bonus, but on the other hand we work twice as much as a full-time employee. [...] Both of us [the full-time employee and the long-term seasonal workers] are keeping the sailing school going, except he's the one who'll get the bonus. That's the problem! (Interview, November 2014)

These inequalities or hierarchies are often euphemized ('it doesn't create tensions'). A few minutes later during the interview, the same instructor actually told me that there is no hierarchy in the water sports centre where he works, citing that she sometimes has lunch with her superior and that she 'dare[s] to tell him anything'. If they arise, complaints could be sanctioned by the employers (who have the power not to hire the worker again for the following season) and badly perceived by colleagues (who see complaining as a bad thing, liable to create a 'negative atmosphere' in the centre). While perceptions of divergences of interests within the professional group may be euphemized, the fact remains that feeling 'sick' of things or like one is being treated unfairly can contribute to a devaluation in the relationship to the work. In this study, I have found that this devaluation process explains the feeling of fatigue at work and is a factor in aspirations to leave the profession.

4.3   Water Sports Without Instructors: Under-Recognized Working and Employment Conditions

Questions pertaining to job instability and the quality of relationships to work do not appear to be much of a concern for the local authorities that promote water sports activities. At best, a simple mechanical link between 'tourism development' and 'better jobs' persists in institutional discourses. Here, for instance, one of the officials in charge of the tourism division at the Regional Council of Pays de la Loire discusses a call for tenders aimed at distributing subsidies for the renovation of buildings in water sports centre:

> I for one think that this kind of scheme can help consolidate tourism activity at the centres, and in the process work towards securing jobs on the sites… The idea is to have a vision of water sports centres that is more economic than purely sport-oriented. […] We do work with them particularly on reaching out to tourist customer bases. […] The goal is essentially economic. It's not about sports. That's not our place. Obviously, it benefits everyone including local athletes, especially them even… But the goal really was to have centres that are in and out of themselves attractive to visitors, right. (Interview, March 2014)

The objective of having water sports centres geared towards commercial activities for tourists, in the name of the benefits this is meant to generate for club members and for salaried workers alike, is explicitly formulated here. The excerpt above express a shared belief that justifies the economic orientations of both water sports centre and local public authorities while ultimately downplaying the question of working and employment conditions. It is in the best interest of local authorities to support and legitimize the touristification of water sports activities insofar as these may become territorial resources (Gumuchian & Pecqueur, 2007), liable to yield symbolic and economic benefits.

The case of the water sports centre of Saint-Jean-de-Monts, a large popular resort on the coast of Vendée,[9] sheds some lights on the connections between the employment question and the successful transformation of a water sports facility into a territorial resource. From the 1950s to the early 1990s, water sports activities in Saint-Jean-de-Monts were fairly undeveloped, and managed by a sports association for mostly non-tourist customers. Activities focused on sport or 'fun', allowing to

---

[9] Saint-Jean-de-Monts has a somewhat older (and ageing) population of around 8800, with a fairly strong presence of craftspeople and shopkeepers. The resort is mostly tourism-oriented, with up to over 100,000 visitors coming in during the summer months. The number of campsites (around 15% of all campsites in Vendée) and the high proportion of secondary residences (66% of the housing stock) reflect the importance of tourism and of the summer season for the town. Source: INSEE, 2014.

spend quality time among peers, were offered. In 1989, a new mayor came into office, which contributed to transforming the management of water sports instruction in the municipality. He intended to boost the image of the town and its infrastructure. This involved retaking control of water sports activities by terminating the association, which had been losing momentum for a few years, and creating a centre staffed by salaried workers under the municipality's more or less direct supervision. The initial goal was to extend the touristic season and to associate the town with water sports. In 1993, a semi-public company (SEM) was created to organize the local tourism offer, which included the water sports centre. The context was conducive to such initiatives, as the French state increasingly delegated the management of tourism and touristic infrastructures to local authorities, encouraging the local interpenetration of the private and public sector (Cousin & Réau, 2016). While, between 1990 and 1994, water sports activities were used for communication purposes, they remained somewhat limited, both in terms of resources granted to the centre and of time, being offered essentially during the summer. By then, stabilizing jobs was not the order of the day. In 1995, the arrival of Charles, the new director of the centre resulted in a change of policy, with a greater focus on job creation. From the beginning, the new director, hired under the government's new youth employment contract scheme, sought to redirect the centre's activities towards tourism, intending in the process to ensure stable employment for himself. One of the major changes he introduced consisted in discontinuing dinghy courses, which were the leading activity for the association's heirs, and replace them by catamaran courses, which are accessible to a wider audience. He contributed to the economic rationalization of the water sports centre and increased activity during the summer months.

> [Charles]: 'In 1995, I started my first career as a sport instructor, with touristic sailing as a goal. I wasn't in it for sport sailing, even if I'd done a little bit of regatta. Tourist sailing, and the goal of living on the coast, and living there year-round, so trying to secure a stable job. It worked out quite well. [...] Today, you'd say I was the director of the centre, at the time you'd barely use base manager. I was starting out, I was on my own in this beautiful facility, and by the time I left in 2004, there were at least four stable jobs. [EG]: Right, you've watched the centre grow!
>
> [Charles]: It's not that I watched it grow, I grew it myself! (Interview, February 2015)

The conversation above marked a significant shift in perspective in my study: the sports instructors are just as involved, if not more, in the production of a supply of tourist sailing activities as those in charge of organizing the supply at the territorial level and the employers/facility managers. This applies particularly well to the cases in which local authorities are directly in charge of managing water sports centres, but also valid for private centres on which local authorities rely to associate territory with water sports. Likewise, employees in associations told me that they felt they were working hard to produce tourist offerings without being supported by their employers' boards, even though they use these offerings to make investments and obtain subsidies. The employees' posture in this configuration can be a subject of tension at work (Falcoz & Walter, 2009). As in other sport worlds, water sports instructors tend to resemble independent entrepreneurs even when they are salaried

(Loirand, 2005), due to the need for them to market their services for their economic survival (and that of their centre).

Thus, even though Charles effectively justified the creation of stable jobs, he contends that he gained little recognition for his work. Water sports were used for communication purposes but actual political efforts to support them remained modest. Operating with a budget he deemed insufficient and facing the semi-public company's refusal to 'upgrade his position' to give him more leeway on management, Charles found himself at loggerheads with his employers and left the company in 2004. He was not replaced by a stable manager; the centre's secretary and administrator first assumed managerial duties, and was succeeded by one of the team's instructors. The latter had little decision-making power over the centre's management, which they thought was poor. Additionally, the building was starting to show signs of wear to an extent that it made working conditions difficult. This was further evidence that for the municipality, water sports activities and the instructors' working conditions were not priority concerns.

The construction of new premises designed to 'showcase water sports', as the October 2007 municipal bulletin put it, began in 2007, in an effort to 'complement the marketing strategy laid out by the semi-public company [by becoming] the focal point of our positioning on "vital emotions, pleasures and experiences" succeeding the image emphasizing the resort's open spaces' (April 2009 municipal bulletin). The construction of the water sports centre was part of a broader policy of renovation of facilities and transformation of the resort's image, which up until that point had been associated with mainstream tourism and safe, family-oriented beaches. In the second half of the 2000s, the library, the multimedia library, the water park, the congress centre (renamed Odysséa) and the tennis courts were renovated or expanded. Building these 'structural facilities', to use the mayor's terms, aimed both at bringing full-time residents and boosting tourism. The new building was fitted with a water recovery system and solar panels, earning it a 'sustainable development ribbon' awarded by mayors' groups, thereby associating with the town with a positive image, beneficial especially to political and institutional actors in the process of justifying their actions (Cousin, 2006). Odysséa and the water sports centre were inaugurated with great fanfare, with both the mayor and the president of the region attending. They were presented as contributing to 'structuring a quality tourism offer' (December 2009 municipal bulletin). Other factors contributed to the construction of the new building. Neighbouring towns had themselves recently renovated their own facilities, forcing the municipality to take part in the competition over local water sports practitioners. Also, and most importantly, the Departmental Council of Vendée and the Regional Council of Pays de la Loire supported the development of water sports as a means to promote the coast at the time. The Saint-Jean-de-Monts water sports centre received 32,000 euros from the region for fleet renewal and 112,500 euros from the department for the buildings.

Despite this influx of resources, the municipality had no precise plan in place for water sports. The new centre director, Xavier, backed by members of the semi-public company, managed to have his working method prevail and to legitimize the place of water sports in the town in the eyes of elected officials. Xavier is a sport sailing instructor. He first managed a water sports equipment shop (specialized in windsurfing) from 1985 to 1995. Then, between 1995 and 2007, he successively managed two water sports centres, largely contributing to their economic development. His experience as a manager was recognized, giving him legitimacy in the eyes of elected officials, but also of the salaried workers he worked with at the centre. He was able to impose his managerial and commercial conceptions, geared towards the creation of tourism-oriented 'products'. He made many investments to renew equipment and introduce new activities, such as kitesurfing, kayak and surfing, allowing the centre to diversify its customer base. Other niche activities were also brought in, such as kite buggying and sea wading. Xavier was thus able to demonstrate that he created stable jobs, bringing distinctive and profitable additions to the local water sports market. This turned out to be worth the company's while: the turnover generated by centre more than doubled over a few years, to the extent that it became one of the most prominent water sports centres in the Pays de la Loire region.

> At first, when I came in, the [company's director] was aware [of the question of development through water sports]; when I asked him whether the municipality was also aware of it, he told me that at the time, the municipality did put resources on the table for renovation, but that there was no specific objective regarding water sports development. Which is no longer the case today, as there has been much progress. First off, the building was in a bad way, it didn't make you feel like promoting it, and then when you don't promote the venue even the workers do not feel committed, you know. There maybe wasn't as much of a team spirit, and as much motivation as there is today. And now, it is showing, including through the people who come to the centre, and they tell us in their satisfaction survey responses [...] You have this positive image even though it used to be less positive, if not negative before, and the local politicians hear about it and they become aware, that, as we've broadened our offer, we're a pillar of tourism in Saint-Jean-de-Monts, right. And the mayor is saying it too. I've heard him say it and tell me, and tell it to other people. (Interview, February 2014).

Xavier, whose legitimacy is recognized both by the company and the municipality, and his team of sports instructors managed to 'assert their activity's contribution to the local economy' (Audinet et al., 2017) and to boost its legitimacy in the local political field (Bourdieu, 2000). While it has generally received support from elected officials and from management, the water sports offer and its development have chiefly depended on the workers, their dispositions to be proactive in their job and their ability to legitimize their choices.

Surfing is a fine example of a 'bottom-up' choice. Saint-Jean-de-Monts is not considered a surfing spot. For experienced surfers, the waves there are too small. The resort has a 'family-friendly' reputation, with risk-free beaches that have little to offer for surfing in technical terms. In 2012, Xavier negotiated the introduction of surfing, which seemed like an eccentric idea, arguing that he could cater to a potential customer base of beginners. He then invested in boards and hired a surfing instructor (on the condition that the latter would also supervise other activities). The instructor in question readily told me that he would look down on the people who surfed in Saint-Jean-de-Monts, but that he quickly understood the potential of a spot with 'safe' waves, 'perfect for children'. The appeal of a stable contract, less dependent on seasonal work and offering better working conditions than the small private schools that dominated the market, led him to accept the offer right away. Surfing was a hit with children and beginner vacationers from the beginning, and immediately turned a profit. Thanks to this success, as well as the introduction of activities for schoolchildren and the creation of a sports association for children by sports instructors from the company (which seconded them for the occasion), in the following years, an open-ended contract position was stabilized, and an eight-month seasonal employee was hired, working mostly on surfing. Surfing ultimately became an important tourist activity in town.

While the introduction of surfing turned out to be a success, regular negotiations between the water sports centre team and the semi-public company's management also sparked tensions. The stabilization of some seasonal jobs, in the form of eight-month contracts, is periodically discussed but never obtained, even though these workers largely contribute to the centre's solid financial performance. There are also lingering conflicts pertaining to the definition of the centre's activity. I observed, for instance, that some sports instructors were outraged when their managers presented a plan to open a snack bar on the premises (serving chips, hot dogs, burgers, etc.) to attract more customers. The instructors believed the idea clashed with the centre's 'spirit', its focus on sports and 'all the efforts that [had] been made to project an image of professionalism'. While anecdotal, the tensions surrounding this snack bar project – which was eventually shelved – mostly reflect the distance between the interest of the semi-public company's employers, looking to attract tourists, and the instructors' relatively inflexible conceptions of their own role and work.

What transpires in the case of Saint-Jean-de-Monts is that the potential to transform the water sports centre into a local resource was actualized through the combination of the presence of workers who were eager to have the centre adopt a market logic and of a conducive political situation. Obviously, whenever it occurs,[10] the form of such a transformation may vary depending on local configurations. While the recognition of the instructors and of their work can be stronger in conducive contexts, as in Saint-Jean-de-Monts were looking to use water sports to promote a new image, it remains in most cases limited. The sport instructors I interviewed regularly pointed out how difficult it is for them to defend their standing and to secure guarantees for the stabilization of their jobs, even in privileged settings like Saint-Jean-de-Monts (characterized by an abundance of financial resources). Léo, for instance, is a sport sailing instructor salaried under six-month seasonal contracts in another well-endowed semi-public centre in Vendée. During the summer season, Léo is entrusted with a management of a site that attracts a sizeable number of tourists. Despite these responsibilities and the fairly significant financial resources at the disposal of his employers, his future remains uncertain. The prospect of securing longer contracts has remained a promise, as seasonality remains strong and continuing to offer competitive sport activities out of season is

---

[10] For instance, an instructor who manages a water sports association in Loire-Atlantique told me that the municipality did not support his activity at all: in his town, tourism focuses on heritage, and the mayor turned his political support to football, probably a more profitable choice to reach out to local voters, who do not particularly favour water sports.

an uncertain gambit. Securing an open-ended contract is a hope in which he tells me he no longer truly believes, since 'all the positions are filled'.

> At the moment, they don't know what they'll be able to offer me next year. They'll probably have a six-month season. But maybe in the off season I could deal with the competition more and he could hire me for eight months, or even nine months. [...] Now eight months, that's getting interesting! It's almost like a real employment contract! (Interview, October 2014)

Even the sports instructors in more stable positions also find it difficult to have their working and employment conditions recognized. Louis, a sport sailing and land sailing instructor, was hired under an open-ended contract by a Loire-Atlantique association in 2003 to boost local water sports activities, in terms both of sport and of tourism, with the municipality's active support. He was successful in that assignment, at the coast of working numerous extra hours (sometimes to the point that he stayed at work overnight during the summer season) and performing management tasks paid below the rate set under collective agreements. After years of practice, he demanded a raise and organizational changes designed to improved working conditions. His employers turned him down systematically, and the situation turned into a conflict. In 2010, Louis felt forced to resign from his job and more broadly to leave the professional group out of disillusionment.

> It was total war, you know. Just nonsense. There was no way out of it except either I left, or they left. And then after a while there was no way around my leaving, because you know… No one gave a shit about me, you see. I was easily replaceable. They had these prominent positions, they were going to lose face locally. And they had power anyway. (Interview, April 2017)

Thus, as in the case of 'surfing without surfers' in the Landes department (Guibert, 2006), the local authorities that publicize water sports may appropriate the profits derived from a territorial resource without always concerning themselves with the workers who produce it. The balance of power tends to be in their favour, as their role is indispensable to the economic activity of sports instructors and by extension to the sustainability of their job (by awarding subsidies, official authorizations, making premises available, promoting activities, etc.), even in cases where these instructors work in private facilities that are not legally subordinate to these authorities.

4.4   Conclusion: The (in)Visibilization of Work and the Future of Water Sports Instructors

The heightened touristification of water sports can yield economic and symbolic resources that are distributed within the market of which water sports centres are a part. While the sport instructors in charge of supervising these activities may find this worth their while to the extent that it justifies their jobs (and the stabilization of these jobs), they are not all on an equal footing on this market. Touristification does not always entail the recognition of their work, although it is necessary to the production of the tourism offer. While touristification to some extent reflects a form of instrumental reality, it also induces values, such as self-sacrifice or passion, that lead to the legitimization of two forms of domination, in the sense of the 'chance that specific commands will be met with obedience on the part of a specifiable group of persons' (Weber, 2019). On the one hand, there is the domination of those who, within the professional group, are in positions of power, which are generally synonymous with stability, over those whose situations are uncertain and who hope to last in the job. On the other, there is the domination of institutions that are partly entitled to define the activities of water sports centres without considering the conditions of those who work there. Thus, touristification contributes to power relations as a result of which aspects of work can be recognized or on the contrary downplayed, or even negated (Krinsky & Simonet, 2012).

These effects of (in)visibilization can impact satisfaction at work in positive and negative ways. The analysis of trajectories of water sports instructors indeed shows that balances between various forms of visibilization and invisibillization, more or less rewarding or degrading, make sustained employment possible. To maintain their position, these instructors must access the possibility of getting recognition for their work without being constantly reminded of their inferior position, and value the positive aspects of the job without being subjected to demeaning quips ('it's not a real job'). These fragile balances are what it takes to reconcile expectations and the reality of working and employment conditions. They are necessary to find a source of 'happiness at work' (Baudelot et al., 2003) and to keep envisioning a lasting future in the job. This is an ambivalent situation because touristification creates expectations – of stability, recognition, good pay – but at the same time unequally distributes the chances of these expectations coming true, which leads to disparities and frustrations that sometimes result in the development of strategies to exit the professional group (Bourdieu, 1974). The challenges and inequalities discussed here are very likely to endure as long as an alternative to tourism as a means of development does not emerge.

References

Audinet, L., Guibert, C., & Sebileau, A. (2017). *Les sports de nature; Une catégorie de l'action politique en question*. Vulaines-sur-Seine.

Baudelot, C., Gollac, M., Bessière, C., Coutant, I., Godechot, O., Serre, D., & Viguier, F. (2003). *Travailler pour être heureux ? Le bonheur et le travail en France*. Fayard.

Bernard, N. (2016). *Géographie du nautisme*. Presses Universitaires de Rennes.

Bourdieu, P. (1974). Avenir de classe et causalité du probable. *Revue Française de Sociologie, 15*(1), 3–42.

Bourdieu, P. (1994). Un acte désintéressé est-il possible ? In *Raisons pratiques. Sur la théorie de l'action* (pp. 149–167).

Bourdieu, P. (1996). La double vérité du travail. *Actes de la Recherche en Sciences Sociales, 114*(1), 89–90.

Bourdieu, P. (2000). *Propos sur le champ politique*. Presses Universitaires de Lyon.

Cartier, M., & Lechien, M.-H. (2012). Vous avez dit « relationnel » ? Comparer des métiers de service peu qualifiés féminins et masculins. *Nouvelles Questions Féministes, 31*(2), 32–48.

Cousin, S. (2006). De l'UNESCO aux villages de Touraine : les enjeux politiques, institutionnels et identitaires du tourisme culturel. *Autrepart, 4*, 15–30.

Cousin, S., & Réau, B. (2016). *Sociologie du tourisme*. La Découverte.

Demazière, D., & Pélage, A. (2001). Mutations de la construction de l'insertion professionnelle. Le cas du dispositif des "emplois jeunes". *Éducation et Sociétés, 7*(1), 81–94.

Denave, S. (2009). Les ruptures professionnelles: analyser les événements au croisement des dispositions individuelles et des contextes. In M. Bessin, C. Bidart, & M. Grossetti (Eds.), *Bifurcations. Les sciences sociales face aux ruptures et à l'événement* (pp. 168–175). La Découverte.

Faguer, J.-P. (2007). Le relationnel comme pratique et comme croyance. *Agone, 37*, 185–203.

Falcoz, M., & Walter, E. (2009). Être salarié dans un club sportif : une posture problématique. *Formation/Emploi, 108*, 25–37.

Faure, J.-M., & Suaud, C. (2009). Privé/public: catégories pratiques ou catégories d'analyse ? Quelques interrogations autour d'une évidence politique appliquée à l'espace des sports. In C. Guibert, G. Loirand, & H. Slimani (Eds.), *Le sport entre public et privé : frontières et porosités* (pp. 265–279). L'Harmattan.

Goffman, E. (1973). La mise en scène de la vie quotidienne, tome 1. In *La présentation de soi*. Minuit.

Guibert, C. (2006). *L'univers du surf et stratégies politiques en Aquitaine*. L'Harmattan.

Guibert, C., & Sébileau, A. (2014). Rester après la saison : l'économie symbolique du néoruralisme balnéaire. *Juris Tourisme, 163*, 31.

Guibert, C., & Slimani, H. (2011). Emplois sportifs et saisonnalités. In *L'économie des activités nautiques : enjeux de cohésion sociale*.

Guillaud, E. (2017). Faire face au contretemps pour faire son temps. *Temporalités* [En ligne], 25. http://journals.openedition.org/temporalites/3685

Gumuchian, H., & Pecqueur, B. (Eds.). (2007). *La ressource territoriale*. Economica.

Hughes, E. C. (1996). Le regard sociologique. In *Essais choisis*. EHESS.

Koebel, M. (2011). Le sport, enjeu identitaire dans l'espace politique local. *Savoir / Agir, 15*, 39–47.

Krinsky, J., & Simonet, M. (2012). Déni de travail: l'invisibilisation du travail aujourd'hui. Introduction. *Sociétés Contemporaines, 87*, 5–23.

Loirand, G. (2005). De la permanence des relations "d'homme à homme" dans le travail d'encadrement sportif. *Cahiers Lillois d'Economie et de Sociologie, 46*, 147–170.

Réau, B., & Poupeau, F. (2007). L'enchantement du monde touristique. *Actes de la Recherche en Sciences Sociales, 170*, 4–13.

Weber, M. (2019). *Economy and society*. Trans. By Keith tribe. Harvard University Press.

Etienne Guillaud is PhD in sociology and is currently a post-doctoral fellow for the CLASMER project, at UBO in Brest (France). He defended a PhD on professional wear process among nautical sports teachers, at the University of Nantes in 2018. His research focuses on working and employment conditions in the leisure, sport and tourism sectors.

Chapter 5
Transformations of Employment and Employment Status, Categorization of Jobs and Competencies: A Legal Perspective on the Organization of the Job Market in the Field of Tourism and Sport in France

François Mandin

Abstract The profession of tourist sports organizer does not exist in France in the order of the categories of law, although it is identified on the socio-economic level. The law knows the professions of tourism organizer and sports educator. However, a close link exists between tourism and sport, especially in so-called nature sports. It is because of the merger of the National Union of Mountain Centers (UNCM) and the French Nautical Union (UNF) that the UCPA (Union of Outdoor Sports Centers), a sports tourism provider, was created in 1965.

Tourism feeds the sporting leisure supply at the same time as sporting leisure feeds the tourist offer. Public land development policies (cycle paths, marinas, etc.) but also employment policies in the field of tourism and sport bear witness to this articulation. The advantages seem numerous: structuring and attractiveness of territories, development of the services and employment market, etc. However, reservations are also made. The market is diversified, fragmented, torn by opposing logics that emerge in at least three places: the offer of services, the offer of training and the offer of employment.

Keywords Regulations · Profession · Conditions of exercise · Diplomas · Competencies · Tourism · Water sports

F. Mandin (✉)
Faculty of Sport Sciences, Nantes University, Nantes, France

Nantes University, CNRS, Law and Social Changes, Nantes, France
e-mail: Francois.Mandin@univ-nantes.fr

© The Author(s), under exclusive license to Springer Nature
Switzerland AG 2023
C. Guibert, B. Réau (eds.), *Employment and Tourism*, SpringerBriefs in
Sociology, https://doi.org/10.1007/978-3-031-31659-3_5

The profession of *animateur sportif touristique* (literally, sport activity leader in tourism) has no legal existence in France even though it is an identified socio-economic reality (Bouchet & Bouhaouala, 2009), whereas *animateur de tourisme* (tourism activity leader) and *éducateur sportif* (sport instructor) are both recognized by the law. Yet, tourism and sport are closely connected, particularly in the so-called outdoor sports (Dubois & Terral, 2011;Langenbach, 2012). The UCPA (Union of Outdoor Sport Centres), a sport tourism provider, was born in 1965 out of the merger of the UNCM (National Union of Mountain Centres) and of the UNF (French Water Sports Union) (Roblot & Verdier, 2015).

Tourism contributes to the provision of leisure sport activities, and leisure sport activities themselves contribute to boosting tourism. Public land use planning policies (cycling paths, marinas, etc.) (Comité Départemental Olympique et Sportif de Vendée, 2016; Bernard, 2017) and employment policies (Karam & Monnereau, 2016) in the fields of tourism and sport reflect this interrelation. This seems mutually beneficial in many ways, in terms of the structuring and economic appeal of territories, of the development of a services and job market, etc. Yet, reservations have also been formulated (Enquêtes, 2018). The market is diversified, scattered, torn between opposite logics in at least three areas of supply: services, training, and employment (Fleuriel, 2016; Chevalier & Pegard, 2016).[1]

The services offered and the statuses of firms vary widely. Tourist and sport services may seem easy to tell apart. Tourism involves a movement of people from one point to another. This trip may include a range of activities that will vary according to the tourist's choices. The tourist offer thus mainly consists in assisting the tourist in the preparation and in the course of his or her trip. Article L. 211–1 of the French tourism code stipulates that a tourist service consists in selling tourist packages, travel services relative to transport, housing, vehicle rental or other travel services that they do not produce themselves. Sport service provision is radically different. It is not defined in the sport code. Applied to supervisory activities, it refers to the supervision of practitioners of a sport in a leisure or competitive setting. These practices may in effect overlap (a hotel stay and a first dive, open-air accommodation and activities for children) or even merge when sport is the purpose of the trip (trekking, sailing courses, scuba-diving trips, etc.). The firms that offer such services are either for-profit companies or non-profits. The former operate in the field of tourism. The for-profit company is a legal form of management that is well-suited to conducting commercial activities and employing salaried workers. These companies are formed for the purpose of generating profits by offering competitive services. Non-profits, called associations in France, are civil organization whose purpose is other than profit. As a result, in the field of sport especially, due to the cost of equipment, they receive government support (from the Ministry of Sports and local authorities) and rely on volunteers. Despite these differences, some overlap can be observed. Sport associations, for instance, partly due to the professionalization of instructors who expand their provision of services to members

---

[1] On the reality and sociological outlines of this market, see Sébastien Fleuriel (2016), Chevalier Vérène and Pégard Olivier (2016).

(diving trips organized by clubs…) or tourist customers (first dives, sailing courses, surfing, kitesurfing, etc.). They can be said to partake in a "commercial" logic, which raises challenges in terms of competition law, fiscal law, accounting and social law. Studies on employment in sport, whenever they address the question of service providers, have evidenced the growing mismatch between their mainly non-profit legal form and their increasingly market-oriented purposes. The decrease of the number of volunteers and the rise of the number of self-employed workers has made the issue even more pressing. In this sense, efforts would need to be focused on the associations to monitor and structure employment in sport. Several schemes exist. Some, like employers' alliances or umbrella companies, pool jobs and in the process cut management costs. The employers' alliance in the "sport and leisure profession" illustrates this trend. Operating under the 1901 French law on associations, it provides employees under a work contract with a single employer and the possibility of securing a full-time open-ended CDI contract, complete with training opportunities. Member employers are spared the task of managing workers, and expenses incurred in managing these pooled jobs are shared by all members. Other schemes pertain to the legal status of companies. The sport sector could for instance adopt the new form of the Collective Interest Cooperative Societies (SCIC). The purpose of these societies is to produce goods and provide services geared towards the collective interest, with a social value *(Loi n° 2001–624 du 17 juillet 2001, titre II ter de la loi 47–1775 du 10 septembre 1947)*. They are subject to the provisions of the Code of Commerce and have a capital. Associations can also create a for-profit subsidiary to optimize their resources; the UCPA is an example of this.

This diversification can also be observed in the provision of training. Training courses for tourism jobs are offered by public institutions that award vocational training certificates (BTS), bachelor's and master's degrees, and private schools, especially business schools, that award bachelor's degrees in business. The supply of training appears balanced both quantitatively and qualitatively. The situation is somewhat difference in the field of sports. Due to the wide variety of disciplines and the safety requirements involved, there are multiple coexisting professional titles. Appendix II-1 (Art. A 212–1) of the Sport Code lists nearly 300 certifications, not including master's degrees and PhDs.[2] Courses are offered by the Ministry of Sports, university sport departments ("Sciences and techniques of physical and sport activities" - STAPS) and sport federations. Some overlap can also be observed here. The National Directory of Professional Certifications (RNCP) classifies the vocational bachelor's programme "Tourism and leisure jobs" and the vocational certificate for youth, popular education and sport (CP JEPS) within the NSF (training specialties nomenclature) category no. 335, entitled "Sport, cultural and leisure activities supervision". This is however the only thing they have in common. The tourism bachelor's programme trains students in management in the field of sport, cultural and leisure activities, whereas the CP JEPS is focused on instruction and supervision jobs. In practice, it would be useful to build bridges between these

---

[2] See the France Stratégie report on sport jobs, p. 67, and the Conseil d'Etat's report.

programmes, in particular for small outdoor accommodation structures looking to provide sport services, but whose volume of activity would warrant one multi-skilled position (management + instruction and supervision) as opposed to two distinct positions for the two types of tasks. The question of employment is, indeed, the third stress point.

Regarding employment, jobs in the tourism and sport sectors appear to be following an identical trajectory, only with a gap of around 10 years (De Lassus, 2008; Enquêtes, 2018). Mechanically, the jobs tend to fit the outlines of the activity sector in which they are embedded. According to the French Centre of Studies and Research on Qualifications (CEREQ) "employment in the sport sector presents characteristics that relate to the very nature of the activity (frequent part-time jobs, fragmented employment, multiple employers and combination of the salaried and independent worker statuses) (De Lassus, 2008; Enquêtes, 2018)." This tends to favour the development of precarious forms of employment: pluriactivity, part-time work, fixed term contracts, etc. (Enquêtes, 2018; Falcoz, 2016; Jolly & Flamand, 2019). This phenomenon is heightened by the seasonal dimension of tourist activities, especially on the coast (Guibert, 2012). Yet, employment policies and labour law do provide tools (state-subsidized contracts, umbrella companies, pooling of jobs) that the actors use.

The "resources" or "tools" offered by law are thus part of the legal organization of the employment and labour market in the tourism and sport sector (Gaudu, 1992). The tourism and sport market is subject to regulations, both regarding the marketing of services and labour. The law serves the freedom of the actors who operate in these tight interconnected markets (business/labour – tourism/sport). These markets cannot exist without the presence of the law (Stanziani, 2004), which is crucial to the regulation and orientation of legal relations that develop as freedoms are used and translated in contractual forms. It is all the more essential as studies conducted in the field, particularly where sport is concerned, show significant disparities that may have adverse effects on employment and by extension on all relations, including commercial ones, in the sport tourism sector (Enquêtes, 2018).

The overall legal picture in the sector may seem surprising. The market of tourist and sport employment is legally structured. Diplomas and training courses are available, at least as per the RNCP directory. Competencies and qualifications are identified. Social partners fulfil their roles. Yet, while employment appears to be up, the development of sport tourism does not seem to be taken into consideration in the field of employment law. It appears that the regulation of the job market, especially in the sport sector, may have structured it but also overly compartmentalized it by introducing diversified training courses whose specialization poses the risk of excessive rigidity. Thus the Conseil d'Etat observed that the fragmentation of certifications, although it should be noted that it is the outcome of a legally valid cooperation process, "may (…) constitute a hindrance to the development of employment, access to sport and the construction of attractive and diversified career paths, even though equivalences exist between certifications providing access to sport instruction jobs(Conseil d'Etat, 2019a, b)."

This hybridization of the provision of tourist and sport services calls for further examination of the structuring of the job market in the sector and its compartmentalization.

5.1 A Legally Structured Market

5.1.1 Guiding Principle

The French job market is structured around the principle of professional freedom, proclaimed by the Allarde Law of 1791. This includes free enterprise (self-employed work) and freedom to work (wage work) (Ferrier, 1997; Lyon-Caen, 2002). This has several consequences. Anyone is free to choose their occupation and the corresponding form of employment (self-employed/wage work). An individual who has chosen to be self-employed may freely pick who they work with; an individual who has chosen to pursue wage work may freely choose their work and their employer. They are entitled to refuse working for an employer. Access to a job should not, therefore, be conditional on the possession of a legally required diploma.

5.1.2 Partial Exceptions

In practice, the exercise of this freedom is tempered by a number of provisions. Some, like the non-competition clause, have contractual roots; others, such as the obligation to hold a qualification permitting access to a profession, have legal roots. The latter obligation characterizes regulated professions.[3]

The tourism sector does not fall under this category. The tourism code does not make access to the profession contingent upon possession of a qualification, meaning that anyone is free to pursue a professional activity in the sector. However,

---

[3] Regulated professions are not defined in French law (Perrin, 2008). In EU law, Directive 2005/36/EC of 7 September 2005 on the recognition of professional qualifications (Art. 3§1, a) defines a regulated profession as "a professional activity or group of professional activities, access to which, the pursuit of which, or one of the modes of pursuit of which is subject, directly or indirectly, by virtue of legislative, regulatory or administrative provisions to the possession of specific professional qualifications; in particular, the use of a professional title limited by legislative, regulatory or administrative provisions to holders of a given professional qualification shall constitute a mode of pursuit." The Court of Justice of the European Union ruled that "a profession cannot be described as regulated when there are in the host Member State no laws, regulations or administrative provisions governing the taking up or pursuit of that profession or of one of its modes of pursuit, even though the only education and training leading to it consists of at least four and a half years of higher-education studies on completion of which a diploma is awarded and, consequently, only persons possessing that higher-education diploma as a rule seek employment in, and pursue, that profession." (CJEU, 1 February 1996, Case C-164/94, Goergios Aranitis v Land Berlin).

this freedom does not guarantee access to employment, which requires demonstrating one's skills to potential employers. For individuals seeking salaried work, the value of a diploma is to guarantee employers that their professional abilities are recognized by the profession, and to be entitled to a number of rights, particularly regarding remuneration, depending on the classifications defined by collective agreements (Caillaud, 2020). For instance, the manager of an outdoor accommodation facility is equally free to hire an employee without a degree and a holder of a vocational bachelor's degree in "tourism and leisure jobs". On the other hand, the same employer, should they be looking for a sport activity leader, will be obligated to hire a holder of the diploma required under law – in this case, the applicable text being the sport code. There is a simple reason for this: the professions of leisure and sport activity leaders are regulated, unlike tourism sector jobs.

This obligation originates in Article L 212–1 of the sport code: "The teaching, monitoring or supervision of a physical or sporting activity or the training of practitioners, against remuneration, as a principal or secondary occupation, on a regular, seasonal or occasional basis, subject to the provisions of paragraph 4 of the present article and Article L 212–1 of the present code, are reserved for the holders of a diploma, vocational qualification or qualification certificate:

- guaranteeing the holder's competence in ensuring the safety of practitioners and third party in the activity under consideration:
- and listed in the national directory of professional certifications under the conditions in Article L 335–6, II of the Education Code".

This legal provision constitutes a breach of professional freedom. Any individual who cannot show evidence of holding one of the qualifications mentioned in Article L. 212–1 of the Sport Code will be unable to supervise such activities against remuneration. This breach is justified by a concern for the safety of practitioners,[4] given the physical risks involved in physical and sport activities.[5] Things would be different if the regulation was aimed at or had the effect of excluding certain individuals,

---

[4] This is precisely the meaning of Article L. 212–1 of the Sport Code, which merely reflects a substantial value to which society seems attached (Conseil d'Etat 2019).

[5] Legislation was first introduced in piecemeal fashion, with regulations pertaining to mountain, climbing and ski-related activities (L. no. 48–267 and 48–269, 18 February1948), then lifeguards (L. no. 51–662, 29 May 1951), judo and jiu-jitsu instructors (L. no. 55–1563, 28 November 1955) and eventually physical and sport educators (L. no. 63–807, 6 August 1963). These disparities then prompted the introduction of a general regulation of the supervision of physical and sporting activities. Article 7 of Law no. 75–988 of 29 October 1975 thus extends the provisions of the Law of 6 August 1963 "to all physical and sporting activities". This law was repealed by the Law of 16 July 1984, which contains specific provisions on training and professions. The latter text was revised twice (L. no. 92–652, 13 July 1992; L. no. 2000–627, 6 July 2000). On physical and sporting activities, see: Pierre J., "Des brevets d'État d'éducateur sportif aux diplômes professionnels: de 1963 à nos jours. Réflexion sur les enjeux et les débats relatifs aux réformes du tronc commun", in Denis Bernardeau Moreau, Cécile Collinet. Les éducateurs sportifs en France depuis 1945. Question sur la professionnalisation. PUR, pp.272, 2009, Des sociétés, 978–2–7535–0974–0. hal-00827951. Regarding skiing, the Court of Cassation found that "access to the profession of ski instructor is, in light of public safety, legitimately made subject to the requirement of a diploma" (Cass. crim. 28 March 2017 ("Tour Operator" "Ski limited"), Légifrance.

for instance on the grounds of their nationality. This is not the case here: EU law, which provides for the freedom to provide services and the freedom of movement of workers, prohibits all forms of discrimination on grounds of nationality, including in sport. The EU directive on the equivalence of qualifications states that "requirements [from Member States] restricting access to a profession or its pursuit to the holders of a specific professional qualification (…) are compatible [if they are, among other things] neither directly nor indirectly discriminatory on the basis of nationality or residence (…)".[6] Both the EU[7] and national[8] case-law are also consistent with this. This is a proportionate breach, which does not introduce a total ban, but makes access to the profession legally conditional on holding a qualification,[9] whereas in tourism this access is "symbolically" conditioned. In both sectors of activity, the job market is structured by a range of diplomas that do not appear to complement each other, despite the close ties between the sectors. This results in a legally compartmentalized market.

5.2 A Legally Compartmentalized Market

Diplomas, in that they certify a competence, have a classifying effect that mechanically leads to the compartmentalization of different job markets. Moving from one market to another indeed requires having one's skills in the latter guaranteed by the relevant diploma, or having them recognized through the process of validating past experience. This effect is heightened in the field of physical and sporting activities due to the legal requirement of a qualification to access the profession of instructor. This compartmentalization raises challenges, especially when the worker must resort to pluriactivity(Delhomme et al., 2018), which is frequent in sport, in response to job insecurity(Fiorelli et al., 2012) and its effects(Lerouge, 2009). In this sense, the objective of ensuring the safety of practitioners, which is the reason for the obligation to hire qualified individuals, appears to contribute to job insecurity in workers.

---

[6] Art. 59.3, Directive 2013/55/EU of the European Parliament and of the Council of 20 November 2013 amending Directive 2005/36/EC on the recognition of professional qualifications and Regulation (EU) No 1024/2012 on administrative cooperation through the Internal Market Information System ('the IMI Regulation').

[7] CJEU 15 Oct 1987, no. 222/86, Union nationale des entraîneurs et cadres techniques professionnels du football (UNECATEF) v Heylens: Rec. CJUE 4097; RFDA 1988. 691, obs. Dubouis.

[8] The Court of Cassation found that "the law provided for the obligation to employ qualified individuals to ensure compliance with required safety measures and the public interest", Cass. crim, 28 March 2017. See also: CE. 17 June 2020 (Syndicat national des professionnels de l'escalade et du canyon, Légifrance), CE 16 December 2019 (Fédération française des moniteurs guides de pêche), Légifrance.

[9] The Court of Cassation found that "(…) the obligation to employ qualified individuals to ensure compliance with required safety measures and the public interest (…) is proportionate to the imperative need to entrust qualified individuals with the safety of skiers" (Cass. crim, 28 March 2017, "Tour Operator", "Ski limited", Légifrance.

5.2.1 From Compartmentalization...

The Sport Code provides for two complementary conditions of access to the profession of physical and sporting activities instructor. The first pertains to the scope of the regulation, and the second to the type of diploma. The diploma is mandatory when the activity falls under the scope of the law and must be supervised by holders of a specific diploma. Holding a diploma in itself is not enough; one has to hold the right diploma.

5.2.1.1 The Scope of the Law Requiring the Possession of a Diploma to Supervise Physical and Sporting Activities

Possession of a diploma is mandatory if the following three conditions are met:

First, the "control" of the conduct of a physical and sporting activity by a practitioner against remuneration. The condition of the "control" of said activity is not explicitly addressed by the text. This is a common denominator to all jobs (or functions) covered by Article L.212–1 of the Sport Code: "teaching, animating (...) supervising a physical or sporting activity or training". They have in common the servicing of individuals (tourists, practitioners) who are placed under the technical authority of an instructor by means of a service provision contract(Mandin, 2004). These jobs are distinguished by the modalities under which these individuals are serviced: supervising (*encadrer*) consists in "commanding" a group of practitioners and ensuring that they are practising in optimal safety conditions; animating (here a literal translation of *animer*) consists in supervising individuals by guiding their activity using a programme; training consists in methodically preparing a sport practitioner for a performance; teaching consists in passing on knowledge pertaining to a physical or sporting activity.

Second, the "control" of practitioners. The mandatory possession of a diploma applies to supervising staff but also to the facility, or physical and sporting activities establishment (APS) that hires the instructors.[10] This is an important point, the subject of litigation that illustrates the interplay of actors and the way in which law and the interpretation of law results in paradoxical situations with questionable economic

---

[10] Article L. 322–1 of the Sport Code states that "No one may operate directly or through a third party an establishment in which physical or sporting activities are practised if they have been convicted of an offence under Article L. 212–9". According to the same code's Article L. 322–2: "Establishments where one or several physical or sporting activities are practised shall offer guarantees of hygiene and safety in compliance with regulation for each type of activity and establishment". Pursuant to the code's Articles L. 322–3 and L. 322–4, the managers of the establishments where one or several such activities are practised must declare their activity to administrative authorities, subject to criminal sanctions. Lastly, according to the code's Article L. 322–5, administrative authorities may deny authorization to open, or close any establishment that fails to comply with the conditions laid down in Articles L. 322–1 and L. 322–2 or that fails to comply with the mandatory insurance provisions laid out in Article L. 321–7.

and social effects. The APS establishment is a category that is specific to sport and has no legal definition. It refers to the place of practice, which makes it possible for a ship or a climbing wall to be considered as an APS establishment.[11] It follows from this that firms that put sport equipment at the disposal of clients (this may for instance consist in renting a windsurf board) are not or should not be subject to the mandatory possession of a diploma, as they cannot be considered primarily as venues where the practice is supervised. While this can be easily understood for a sports shop that rents skis, the situation is more ambiguous when the equipment being rented is located on a firm's premises (for instance, an outdoor accommodation facility) on the seaside or riverside. The Sport Code makes provisions for such situations but only where classified establishments that fall under tourism regulation are concerned. Article L. 212–4 of the Sport Code holds that: "The provision of equipment to practitioners, or, outside of the case of activities practised in a specific environment, the facilitation of the practice of an activity within a classified establishment falling under tourism regulation shall not be considered in the same way as the functions referred to in Paragraph 1 of article L. 212-1." This mention was introduced upon the request of hospitality industry professionals who considered the law as "significantly out of proportion and disadvantageous vis-à-vis foreign competition".[12] However, by limiting the obligation to the sole activities practised "within a classified establishment falling under tourism regulation" without taking into consideration the entire range of characteristics of sporting activities involving the provision of equipment, lawmakers failed to resolve the matter. In the case law, sporting activities involving the provision of equipment practised outside of a classified establishment falling under tourism regulation have been equated with supervisory functions of physical and sporting activities: "Recognizing that if the mere provision, by way of a sale, loan or rental, of equipment needed for a physical or sporting practice is not enough to characterize an establishment in which physical and sporting activities in the sense of the aforementioned provisions are practised, in the case of sporting or physical activities practised outside of enclosed spaces, the

---

[11] "An APS establishment is defined as any entity that organizes the practice of a physical or sporting activity. A range of parameters must be met for such an entity to be recognized as such: fixed or mobile sporting equipment (boats, horses, paragliders, etc.), a physical or sporting activity" and a certain duration "which may be a few months (…) or regular but discontinuous" (*Instruction n° 94–049 JS* du 7 mars 1994) (Wagner, 1990; Cass. crim., 2008; Mandin, 2009).

[12] "The topic of supervisory jobs in the tourism sector, and particularly in the leisure hospitality industry, which has been the subject of many parliamentary interventions, directly relates to Article 43 of the Law of 16 July 1984 and the mandatory diploma requirement contained therein (…). The Minister of Sports, Mr. Jean-François Lamour, and the Secretary of State for Tourism, Mr. Léon Bertrand, have thus been led to adopt a joint position regarding the scope of Article 37 of the law on sport. The press release thus states that 'In order to respond to the practical challenges reported by tourism professionals concerned by the implementation of standards they considered overly restrictive, the two ministers have had the occasion to mention that the mere provision of equipment or facilitation of activities, with or without the involvement of staff, would be manifestly excluded from the scope of the new regulation, particularly in establishments operating in the tourism sector, as long as staff did not perform teaching, animation, training or supervision activities. This applies for instance to the organization of a contest or tournament'" (Depierre, 2002).

individual who should be considered as the manager of such an establishment shall be the person who, finding himself in close proximity to the place of practice, organizes a sporting practice within a limited perimeter by providing practitioners with the necessary equipment, and in addition to doing so, also provides instructions, advice or information for the purpose of preventing the risks inherent in the activity, even if he does not offer teaching, animation or supervision services by making authorized staff available to practitioners for the entire duration of the practice"(Conseil d'Etat, 2010). This is a debatable reasoning: it is difficult to understand why the place of practice is the basis for applying Sport Code provisions pertaining to supervision of activities when the renter does not perform such activities. How, then, can this decision be justified? It can be argued that Article L. 212–4 of the Sport Code constitutes an exception and that the safety-oriented rationale of the legal framework leads to extending it to rental firms that may potentially offer supervision services. But the framework does not make sense. Either the firm's activity is the provision of sporting equipment, which is not similar to the supervisory functions for which the diploma is mandatory, or it is supervision, in which case it is subject to the professional regulation. It can also be both, in which case it is also subject to the professional regulation. The framework in place creates confusion and legal insecurity that could hinder the development of equipment provision activities, and supervision activities through a rebound effect. Difficulties of a somewhat similar nature have arisen regarding the meaning of the "physical and sporting activities" condition.

Physical and sporting activities – Possession of a diploma is mandatory when physical and sporting activities are offered. Although it is central, this condition is not defined by the law. It has been interpreted broadly in the criminal case-law. Physical and sporting activities, without "physical and sporting" being defined, are recreational and competitive practices. The following definition from the Criminal Chamber of the Court of Cassation, regarding paragliding, suggests that the activities in question are those that require muscle strength and have an emotional component: "the paraglider is an aircraft, an ultra-light glider that takes off using the muscle strength of its pilot; they note that two-seater flight requires from the passenger, who is suspended by straps, active cooperation upon take-off and landing, as well as sufficient emotional regulation, and in some cases participation in steering during the flight; they infer from this that paragliding constitutes a sporting activity for the passenger and the supervision of said activity for the pilot, and is as such subject to the provisions regulating the teaching against remuneration of physical and sporting activities on which the claims are based; they add that the application of rules resulting from the Civil Aviation Code is not an obstacle to the application of the regulation on sport"(Cass. Crim., 2001). Litigation on this matter reflects both the value of the legal framework, in that it protects practitioners, but also its limitations, as in some activities such as paragliding or sailing, service providers must cumulatively comply with the provisions of sport law and civil aviation law for paragliding, and maritime law for recreational sea fishing.[13]

Remuneration – The remuneration condition is the most straightforward on paper. The mandatory diploma requirement pertains solely to services supervised by an individual who is remunerated to this effect, whether they are self-employed or salaried. Individuals performing unpaid supervisory activities do not have to possess a diploma. This exclusion of unpaid work is inconsistent in light of security requirements, especially as volunteers make up a significant fraction of supervisory staff. This exclusion is the result of the introduction of the term "remuneration" following the concerns voiced by sport organizations. The mandatory diploma requirement, according to their representatives, would have depleted the volunteer workforce.[14]

Yet, sport organizations have developed their own training programmes for volunteers, who now compete with instructors holding a title that allows them to work against remuneration. In response, the national confederation of sport instructors, sport and animation employees (CNES) filed claims against four federations [French federation of mountaineering and climbing (complaint no. 14–26.531) – French sailing federation (complaint no. 14–26.532) – French field hockey federation (complaint no. 14–26,530) – French gymnastics federation (complaint no. 14–26,529)] for the purpose of prohibiting them from using the designations of professor, instructor, educator, trainer or animator of a physical or sporting activity in the titles of federal diplomas that do not entitle holders to have a remunerated practice teaching, animating or supervising a physical or sporting activity or training practitioners. The confederation argued that these federations could not use such designations for training programs that only entitled individuals to volunteer practice. In its four rulings of 17 December 2015, the Court of Cassation did not concur. It found that the scope of Article L. 212–8 of the Sport Code is limited to the teaching against remuneration of a physical or sporting activity. The offence of using protected titles provided for in Article L. 212–8 of the Sport Code[15] applies solely to activities requiring a qualification and therefore only activities of teaching against remuneration, except those carried out on a volunteer basis. Its purpose is not to restrict the use of such titles to supervision against remuneration and by extension to prohibit volunteers from using them. This is interesting litigation in that it characterizes the prevailing logics at work in this market. First, there is a logic based on identity and economic protection: titles indicate the possession of a competence that

---

[13] Regarding the extension of the operation of a professional fishing vessel to recreational sport activities (recreational fishing, diving) (Cass. Crim., 2008; Mandin, 2009).

[14] See Loi du 1er août 2003, J. Bordas, Rapp. AN n° 246 sur le projet de loi relatif à l'organisation des activités physiques et sportives.

[15] "Any individual performing one of the functions of professor, instructor, educator, trainer or animator of a physical or sporting activity or using these title or any other similar title without possessing the required qualification under Article L. 212–1 or performing their activity in breach of Article L. 212–7 without having passed the tests required by the administrative authority (…) shall be liable to a prison term of one year and a fine of 15,000 euros." (C. Sport, Art. L. 212–8, 1).

allows their holders to work professionally on the sport services market. When they use these titles for their volunteers, the federations create confusion on the sport services market by suggesting that volunteers and professionals have equal competences. Then, there is a logic based on safety and the criminal protection of the market: supervision against remuneration is prohibited for any individual that does not possess one of the titles listed in Article L. 212–1 of the Sport Code. Any individual who contravenes this obligation, either by working or by making someone else work, may incur criminal penalties.[16] This applies for instance to the holder of an ultralight instruction diploma if he supervises the practice of paragliding, which is a distinct activity. In doing so, he is committing a criminal offence because even if he has knowledge of air navigation, which might be considered to be sufficient, he does not possess the state certificate for free flight that entitles its holders to supervising paragliding activities against remuneration.[17]

5.2.1.2   Diplomas Permitting Access to Supervision against Remuneration

Article L. 212–1 of the Sport Code states that individuals intending to "supervise" physical and sporting activities against remuneration must possess one of the three following titles: "a diploma, a title for professional purposes or a qualification certificate. Supervision against remuneration is prohibited for anyone who does not possess one of these three titles. The titles must fulfil two conditions. The first is substantive: the diploma must "[guarantee] the competence of its holder regarding the safety of practitioners and third parties in the activity under consideration". The second is procedural: the title must be "listed in the national directory of professional certifications under the conditions in Article L 335-6, II of the Education Code".

Typology of Diplomas and Titles, with an Illustration in the Field
of Recreational Water Activities

Diplomas are awarded by the Ministry of Sports, sport science faculties and the relevant sport federations. These diplomas are organized hierarchically and attest to degrees in the specialization of advanced skills ranging from animation to the training of top-level athletes. These skills are exclusively sport-related, even if, for instance, the "certificate of professional aptitude for technical assistant animation work" (BAPAAT) concerns socio-cultural activities.

The latter certificate, which has since 2019 been replaced by the CP JEPS of equivalent level (III-CAP-BEP) constitutes the first level of professional qualification for the animation and supervision of physical and socio-cultural activities. This diploma allows its holders to perform animation functions under a manager who

---

[16] C. Sport, Art. L. 212–8, 1.
[17] Crim. 14 déc. 2004, no 04–82.401.

holds a higher-level qualification (BPJEPS, DEJEPS or DESJEPS). There is no specific qualification to the field of water recreation activities and sports.

The "professional certificate of youth, popular education and sport" (BPJEPS) constitutes the second level. It indicates that the holder is able to work autonomously as an animator in a specialty area – disciplinary, pluridisciplinary, or focused on a particular sub-field within the specialty area. In the example of sailing, there are six specialty areas. These include the BPJEPS in the "sport educator" specialty with a major in "mono- and multi-hull sailing within six nautical miles of the coast", which allows its holders to supervise and animate activities of introduction to sailing and training on all types of boats, and the BPJEPS in the "sport educator" specialty with a major in "sailboat cruising within 200 nautical miles of the coast" which allows its holders to supervise and animate activities of introduction to and training for sailboat cruising.

The state diploma for youth, popular education and sport indicates that the holder possesses the skills necessarily for working as a coordinator or trainer in a particular sport. The DEJEPS is awarded for the "specialty area" of "advanced sport" with a major in "sailing", allowing it holders to teach, animate, supervise or train in advanced sailing activities.

Lastly, the higher state diploma for youth, popular education and sport[18] (DESJEPS) in the "sport performance" specialty is the highest diploma. It indicates that the holder possesses the skills to work as a project manager, facility manager or team manager in a particular field. This diploma also exists in the field of sailing – the DESJEPS, in the "sport performance" specialty with a major in "sailing".

Sport science (STAPS) faculties also award diplomas that allow their holders to supervise activities against remuneration, like the holders of a youth and sport diploma. First, the bachelor's degrees with a major in "STAPS: education and motricity" and "sports training" allow their holders to "supervise, teach or animate physical or sporting activities at a beginner, intermediate or recreational level". These diplomas are not sufficient to supervise a particular sport autonomously; a mention of the discipline must be included in the diploma. Regarding sailing, the bachelor's degree "STAPS: sport training" allows its holder to supervise sailing activities for the purposes of performance improvement and personal development only when the mention of "sailing" has been added.

Lastly, sport federations can also award professional qualification certificates (CQP). A distinction should be made between titles awarded to unpaid and professional workers. The former, often called federal certificates, allow their holders to supervise activities without being remunerated. The latter, which fall under the CQP

---

[18] This is a "higher state diploma listed in level II of the national directory of professional certifications pursuant to Article L. 335-6 of the Education Code. It certifies the acquisition of a qualification to practice a professional activity of technical expertise and management for educational purposes in physical, sporting, socio-educational or cultural areas of activity" (C. Sport, Art. D. 212–51). This diploma is "awarded for the 'sport performance" specialty or the "socio-educational or cultural animation" specialty, with a major in a specific disciplinary or professional field" (C. Sport, Art. D 212–52).

category, are also awarded by federations, but unlike the federal certificates, they allow their holders to work against remuneration. These certificates are listed in the Sport Code. In the case of sailing, there is a "sailing initiator" CQP, which allows its holders to perform activities of animation in and introduction to sailing under the authority of a BPJEPS holder.

Substantial and Procedural Characteristics of Diplomas and Titles

Substantial characteristic: Safety and the disciplinary specialization of skills.

The diploma must ensure "the holder's competence in ensuring the safety of practitioners". Here the term "safety" must be understood globally. The holder of the diploma received a training that did not exclusively focus on safety, but on knowledge of the practice, its practitioners and its conditions. This knowledge is precisely what is supposed to ensure the safety of practitioners of a particular sport discipline. This disciplinary logic results in the gradual specialization of skills. The first level exclusively attests to a crosscutting skill in socio-cultural animation, exclusively in the field of sport. At this level, sport practice is a means to perform animation activities. The supervision of individuals, regardless of the sport discipline in question, can, outside of the activities that fall under specific regulations due to the particular risks they entail (such as diving or skiing), be performed by holders of a diploma in "sport" that is not specifically specialized in the supervision of a sport discipline. These include:

- an academic diploma, such as th
- e bachelor's in sciences and techniques of physical and sporting activities, or education and motricity, with a major in sciences and techniques of physical and sporting activities.
- a professional title in "youth and sport", such as "animator/technician in physical activities for all", BPJEPS, with the "physical activities for all" specialty mention. The conditions and limitations of the use of these diplomas are laid down in Appendix II-1 (Art. A212-1) of the Sport Code.

However, this crosscutting skill is insufficient to supervise practitioners in a specific discipline. This applies for instance to jobs that consist in supervising sport disciplines such as water sports, which also involve contact with tourists. Businesses that are eager to offer quality supervision will hire an instructor who holds the required diploma. They may be all the more keen to do this as there is an economic risk involved in entrusting the supervision of a particular discipline to an instructor who holds a generalist diploma, under the authority of an instructor with a diploma in that discipline, as he or she may not have sufficiently detailed knowledge of that practice. There is also a criminal risk involved in not employing a person who holds the title required by law.[19]

Procedural characteristics: Registering diplomas in the national RNCP directory.

Any individual seeking to supervise a practice against remuneration must hold a diploma or title that is "listed in the national directory of professional certifications". This directory contains fact sheets on each course. The creation, revision or suppression of diplomas and titles for professional purposes and their reference frameworks is done as part of a tripartite decision-making process involving all stakeholders: the state (through its training bodies), businesses and unions (C. trav., art. L. 6113–1, L. 6113–5 à L. 6113–9; art. L. 6113–3 et L. 6113–4). These fact sheets contain details on the skills acquired and the jobs that can be performed in France: "The professional certifications listed in the national directory of professional certifications allow for the validation of acquired skills and knowledge required for the pursuit of professional activities. They are defined in particular by an activity reference framework that describes work situations and activities performed, the trades and jobs that can be pursued, a skill reference framework that identifies the related skills and knowledge, including crosscutting ones, and an evaluation reference framework that defines the criteria and the means of evaluation of learning outcomes" (C. trav., art. L. 6113–1).

This system of certification and registration of certifications is expected to guarantee the safety of practitioners. Any individual using a non-registered title or not possessing a title cannot engage in the profession of sport instructor. The scope of the work of the diploma's holder is also strictly limited. The diploma's holder cannot be employed to pursue supervisory functions other than those the diploma entitles him to. For instance, a holder of a BPJEPS with a specialty in "nautical activities" and a monovalent major in "sailing" is only allowed to "supervise and animate activities of initiation to sailing, including the first levels of competition in sailing", and is barred from supervising all other activities. He or she will not be allowed to autonomously supervise multi-hull, dinghy, cruising activities beyond 12 nautical miles of the shore. Anyone intending to supervise activities beyond 12 nautical miles of the shore must possess a "cruising" specialty certificate (C. Sport, Annexe I-0-1 - art. A114–3).

5.2.2   ...To Decompartmentalization

The combination of a security-based rationale and of specialization has led to a compartmentalization within sports professions and in the relations of these professions with others, especially in tourism and professional pleasure sailing. The legal system of equivalences constitutes a first avenue to decompartmentalize the market. The second consists in relaxing or removing the mandatory diploma requirement.

---

[19] It has for instance been ruled that "the practice of paragliding (...) is a distinct activity that requires the possession of a state certificate for free flight" (Crim. 14 déc. 2004, no 04–82.401).

5.2.2.1 Equivalences: A Relative Decompartmentalization

Equivalence is a process that consists in recognizing an identical value to French diplomas or academic titles of various origins or to foreign and French diplomas or academic titles. Equivalences can therefore be national or European.

* National equivalences – In the field of sport, French law provides for equivalences in sailing. Even though sailing instructors and sailors have knowledge of sailing, the diplomas they are awarded in their respective activity sectors do not allow them to work in an area other than the one relative to their qualification. A sailor, for instance, whose profession is regulated,[20] as the profession of sport instructor is, cannot teach sailing, and a sailing instructor cannot work on a professional pleasure sailing boat. This prohibition is justified on the grounds that the respective diplomas, even though they both pertain to activities that take place on the sea, are not awarded for the same skills. A sailing instructor is primarily the instructor of a physical and sporting activity that happens to be sailing. Here the core activity is the supervision of one or several individuals eager to acquire or develop skills in a sport discipline. A sailor, on the other hand, is qualified to perform an activity that relates to the operation of a boat.

However, the precariousness of jobs, particularly in the youth and sports sector, and the similarity of the respective required skills have led line ministries to set up an equivalence system between two sets of diplomas: first, the Captain 200 sailing certificate; second, the state diploma for youth, popular education and sport with a specialty in "advanced sport" and a major in "sailing", and the professional certificate of youth, popular education with a specialty in "nautical activities". The nautical certificate of initiation is awarded by equivalence to holders of the first degree state educator certificate with a major in "sailing", of the "cruising" specialization certificate, of the "nautical activities" specialty, of the state diploma for youth, popular education and sport with a specialty in "advanced sport" and a major in "sailing", and of the higher state diploma for youth, popular education and sport with a specialty in "sport performance".[21] Conversely, holders of the Captain 200 sailing certificate may obtain equivalences for certificates of skills pertaining to the supervision of practitioners and the safety of navigation.[22]

---

[20] "No one may pursue the occupation of sailor if they do not possess the professional maritime training titles and qualifications corresponding to the capacities they are expected to have and the functions they may be required to perform on board the vessel." (C. transp., art. L. 5521–2 I)

[21] Arrêté du 5 juin 2013 relatif aux modalités de délivrance par équivalence du certificat d'initiation nautique et du brevet de capitaine 200 voile délivrés par le ministre chargé de la mer aux titulaires de certains titres de formation professionnelle délivrés par le ministre chargé des sports.

[22] Arrêté du 3 juillet 2013 portant équivalence entre le diplôme d'Etat de la jeunesse, de l'éducation populaire et du sport spécialité « perfectionnement sportif », mention « voile », et le brevet de capitaine 200 voile délivré par le ministre chargé de la mer; Arrêté du 3 juillet 2013 portant équivalence entre le brevet professionnel de la jeunesse, de l'éducation populaire et du sport spécialité « activités nautiques » et le brevet de capitaine 200 voile délivré par le ministre chargé de la mer.

An identical approach could be considered for those tourism diplomas and sport diplomas that pertain to sport, culture and leisure activities (no. 335 in the NSF training specialties nomenclature). Given the requirements of sport specialties, this would require introducing an equivalence at CP JPS level, or introducing a sport specialty.

- European equivalences – Equivalences are not limited to the national level. Under EU law, in accordance with the principle of the freedom to provide services and the free movement of workers, it is unlawful to deny employment to an EU national on the grounds of their nationality. To facilitate the implementation of said principle, EU law has set up an equivalence system for European diplomas, allowing any national of an EU member state to offer their services or to answer a job offer on the territory of a member state. The Sport Code according holds that: "Functions listed in paragraph 1 of Article L. 212-1 may be performed within the national territory by nationals of European Union member states or of state parties to the Agreement on the European Economic Area who are qualify to perform them in one of those states. These functions may also be performed, on a temporary and occasional basis, by any national legally established in a European Union member state or in a state party to the Agreement on the European Economic Area. However, when the activity in question or the training leading to said activity is not regulated in the state in which it is established, the service provider must have performed the activity in one or several European Union member states or parties to the Agreement on the European Economic Area on a full-time basis for at least a year or on a part-time basis for an equivalent total duration during the 10 years prior to the provision of the service."[23]

---

[23] C. Sport, Art. L. 212–7. On the ski sector: it has been ruled that the so-called "Services" directive 2006/123/EC allows a member state within which the provider travels to provide a service to impose requirements concerning the provision of the service whenever justified for reasons of public policy, public security, public health or protection of the environment. It follows that the activity of "ski guide" [which] is performed in mountainous areas, a specific environment that poses special risks requiring the intervention of professionals with in-depth knowledge of the mountain environment and its risks to ensure that clients practice under optimal rules (...). [shall not be performed by the] ski instructor of UK nationality] who does not have the qualification necessary for the pursuit of the profession of ski instructor, access to which is open without discrimination to all European Union nationals through the review or recognition of qualifications" (Cass. crim., 13 juin 2017, "Ski club Great Britain ", non publié, Légifrance). On diving: it has been ruled that the equivalence shall not be granted due to the objective differences between diplomas considering that the foreign diploma "entitles only to supervising underwater diving on a volunteer basis" (...) and that the "OWSI" is awarded by an international recreational underwater diving training organization, whereas the state certificate confers on its holder the qualification necessary for the teaching of diving in all its forms" (Cour administrative d'appel de Paris 3 avril 2014).

5.2.2.2 Liberalization

The other option, which is more absolute – perhaps too absolute – consists in removing the mandatory diploma requirement. This is technically easy: it basically requires changing the law. Regarding the substance of the matter, the situation is more complex. While liberalization removes a breach of professional freedom, its potential effects on the increase in sports accidents should be established. This is no easy task. Indeed, considering the mandatory diploma requirement was introduced for safety reasons, it would be logical to assume that removing it would result in increasing numbers of accidents. Yet, this requires being able to demonstrate that the mandatory diploma requirement has reduced the number of accidents through training. The law would then be considered virtuous insofar as it promotes training courses. But how then are we to explain why the mandatory diploma requirement has not worked to the detriment of volunteers? As far as we can observe, the number of accidents has not increased along with the presence of volunteers, which is evidence that the key factor is not the legal mandatory requirement by the training of individuals. In this sense, service providers, if only to ensure the longevity of their activity, have no interest in entrusting the supervision of practices to unqualified individuals. In the event of an accident, they would be contractually liable for failure to fulfil their obligation to ensure the safety of practitioners. In this respect, not being a holder of the required diploma has no mechanical incidence on the practitioner's accident[24] and the service provider's liability.

It appears, however, that the diploma requirement on safety grounds will not be reconsidered. In its latest report, the Conseil d'État has noted that "the safety of sport practices is a constant preoccupation for the state and a core matter for sport law".[25] Perhaps the middle road option of simplification suggested by the Conseil d'Etat should be pursued: "The fragmentation [of the certifications] reflects the great diversity of disciplines and of the conditions of their practice, for the legitimate purposes of qualification of educators and safety of practices. It may however constitute a hindrance to the growth of employment, to access to sport, and to the construction of attractive and diversified professional trajectories, although equivalences exist between certifications for access to sport supervision jobs. Additionally, the current system does not define any criteria for mandatory qualification according to the activity. Yet, it happens not unfrequently that sport science STAPS graduates are forced, having completed their university training, to attend another training course leading to a lower-level diploma in order to be entitled to employment in the supervision of sport activities in an association. It is therefore necessary to review the organization of training courses and diplomas to better combine the intangible

---

[24] Civ. 1, 2 Feb. 1989. It has been found that the organizer of a mountain bike ride could not be liable for a practitioner's fall on the grounds first "that the chosen route for the ride did not present any particular challenges and had already been taken [by the victim], and second that the fact that the mountain bike ride organizer was not a holder of the required diplomas had no incidence on Mr. Y.'s fall..., [the judge has in accordance with the law] justified his ruling".

[25] p. 97.

requirements of the protection of practitioners' safety and the objectives of professional mobility and employment growth. Under this new architecture, a set of competences and qualifications allowing diploma holders to bring managerial skills to sports associations and firms should be identified. This must be supported by a consideration of the professional trajectories of sport educators, in order to offer them stimulating and appropriate career prospects" (Conseil d'Etat, 2019a, b).

5.3 Conclusion

The regulation of sports coaching diplomas originates from the protection of the participants. Thus, this regulation has specified the conditions of access and exercise of the profession, by attesting that the diploma holder has undergone training adapted to his job. This training allows him to practice in complete safety. However, the regulations in force do not designate a single training institution such as the university, as is the case in the medical field, to train instructors. Different institutes can now issue qualifications and diplomas to train instructors. The difficulties encountered are due to the fact that the content and the number of hours are not the same or that there are no equivalences between the training courses provided by these different organizations. These difficulties are not specific to the tourism and nautical sector. They can also be found, for example, in the field of diving, about the diploma of the French Federation of Underwater Studies and Sports (FFESSM) and the Professional Association of Diving Instructors (PADI). The result is a complex legal framework which, according to the French Council of State, should be "simplified (…) in order to offer sports educators attractive career paths and to enhance and develop the jobs concerned".

This simplification is even more necessary as European regulations advocate the equivalence of European training and diplomas, including in the field of physical and sports activities (Code du Sport, Art. L. 212–7). However, this simplification will only have a real effect if the regulations standardize the content and the number of training hours. This presupposes a "consensus" among both training organizations and practicing professionals.

References

Bernard, N. (2017). Nautisme et tourisme: une convergence au bénéfice des territoires. *Études Caribéennes[Online], 36.* https://doi.org/10.4000/etudescaribeennes.10505
Bouchet, P., & Bouhaouala, M. (2009). Tourisme sportif. Un essai de définition socio-économique. *Téoros. Revue de Recherche en Tourisme, 28*(28–2) http://journals.openedition.org/teoros/438
Caillaud, P. (2020). *Le diplôme. Répertoire de droit du travail.* Dalloz.
Cass. Crim. (2001 march). Bulletin criminel 2001. *Cass. Crim., 76,* 250.
Cass. Crim. (2008). Bulletin criminel 2008. Obs, (204).

Chevalier, V., & Pegard, O. (2016). L'emploi sportif: fabrique d'une illusion, fabrique à illusions. *Marché et Organisations, 27*(3), 15–29. https://doi.org/10.3917/maorg.027.0015

Comité Départemental Olympique et Sportif de Vendée. (2016). Schéma du développement du sport en Vendée. Retrieved 9 Sept 2020, from http://www.cdos85.fr/wp-content/uploads/2016/12/Sch%C3%A9ma-de-d%C3%A9veloppement-du-sport_Vend%C3%A9e-.pdf

Conseil d'Etat. (2010). no. 330614. *Légifrance.*

d'Etat, C. (2019a). *Le sport: quelle politique publique?* Les rapports du Conseil d'État.

d'Etat, C. (2019b). *Le sport: quelle politique publique?* (p. 97). Les rapports du Conseil d'État.

De Lassus, I. (2008). *Les métiers du tourisme: tendances et besoins émergents.* Cereq Centre. http://cdt64.media.tourinsoft.eu/upload/Les-metiers-du-tourisme.pdf

Delhomme, I., Deroin, V., & Insee. (2018). *Au cœur du sport, une forte progression des indépendants à côté des associations.* Insee. https://www.insee.fr/fr/statistiques/3650807#titre-bloc-8

Depierre, B. (2002). *Rapport fait au nom de la commission des affaires culturelles, familiales et sociales sur la proposition de loi adoptée par le sénat modifiant l'article 43 de la loi n° 84–610 du 16 juillet 1984* relative à l'organisation et à la promotion des activités physiques et sportives. Assemblée Nationale. From http://www.assemblee-nationale.fr/12/rapports/r0465.asp#P133_15334

Dubois, F., & Terral, P. (2011). De l'amateur sportif au dirigeant d'une petite entreprise. Le tourisme sportif de pleine nature. *Travail et Emploi, 126,* 35–44.

Enquêtes, C. (2018). *Regards croisés sur le secteur des activités sportives et le métier d'éducateur sportif* (p. 2). Portrait Statistique Emploi Formation.

Falcoz, M. (2016). Emplois sportifs, emplois pérennes, emplois précaires. *Marché et Organisations, 3,* 31–44. https://doi.org/10.3917/maorg.027.0031

Ferrier, D. (1997). La liberté du commerce et de l'industrie. In R. Cabrillac, M.-A. Frison-Roche, & T. Revet (Eds.), *Droit et libertés fondamentaux* (4th ed., pp. 504–509). Dalloz.

Fiorelli, C., Tallon, H., Dufour, A., Maïzi, P. M., Massein, G., Pigache, M., & Cadier, C. (2012). La pluriactivité au risque de la précarité: singularités des rapports au travail et à l'emploi dans les activités rurales. Une étude menée en Languedoc-Roussillon. *In Symposium PSDR" Les chemins du développement territorial"* (pp. 26).

Fleuriel, S. (2016). L'autre marché du travail et de l'emploi sportifs. *Marché et Organisations, 27,* 11–14. https://doi.org/10.3917/maorg.027.0011

Gaudu, F. (1992). *L'organisation juridique du marché du travail* (p. 941). Droit Social.

Guibert, C. (2012). Les effets de la saisonnalité touristique sur l'emploi des moniteurs de sports nautiques dans le département des Landes. *Norois. Environnement, Aménagement, Société, 223,* 77–92. https://doi.org/10.4000/norois.4213

Jolly, C., & Flamand, J. (2019). Droits sociaux et statuts d'emploi: une cartographie des métiers. *Regards, 55*(1), 39–52. https://doi.org/10.3917/regar.055.0039

Karam, P., & Monnereau, R. (2016). *Evaluation des dispositifs de soutien a l'emploi dans le champ du sport.* Report no 2016-M-03. Ministère des Droits de la Ville, de la Jeunesse et des Sports, Inspection Générale de la Jeunesse et des Sports. France.

Langenbach, M. (2012). *Le marché du tourisme sportif de nature dans les systèmes territoriaux des espaces touristiques et ruraux: l'exemple de l'Ardèche.* (Doctoral dissertation,. Université de Grenoble.

Lerouge, L. (2009). Les effets de la précarité du travail sur la santé: le droit du travail peut-il s' en saisir?. *Perspectives Interdisciplinaires sur le Travail et la Santé* (11–1). https://doi.org/10.4000/pistes.2306.

Lyon-Caen, A. (2002). Le droit du travail et la liberté d'entreprendre. *Droit social, 3,* 258–263.

Mandin, F. (2004). Obligation légale du diplôme relatif à l'encadrement contre rémunération des activités physiques ou sportives. *Travail et Protection Sociale,* (7).

Mandin, F. (2009). L'application du « droit du sport » au navire exploité pour des activités de plongée et de pêche. *DMF,* (701).

Perrin, A. (2008). *Les professions réglementées* (p. 11). Droit Administratif.

Roblot, A., & Verdier, L. (2015). *50 ans d'histoire avec les jeunes: la saga ucpa.* UCPA Association. Retrieved 9 Sept 2020, from https://www.ucpa.asso.fr/accueil/actualites/actualite/50-ans-d-histoire-avec-les-jeunes-la-saga-ucpa

Stanziani, A. (2004). Comment le droit définit le marché. *Revue d'Histoire Moderne Contemporaine, 513*(3), 199–203. https://doi.org/10.3917/rhmc.513.0199

Wagner, E. (1990). L'exploitation contre rémunération d'un établissement d'APS au regard de la loi du 16 juillet 1984. *RJ Eco, 14*, 3.

François Mandin, Director of the Center for Maritime and Oceanic Law (CDMO). HDR lecturer, Sciences and techniques of physical and sports activities, Nantes university.

Chapter 6
The Space of "Social Tourism". Organizations, Mobilizations and Labour

Francis Lebon

Abstract The notion of "social tourism", which took off in France during the Liberation, reached its peak in the early 1970s. Since then, "social tourism" has lost its importance. It is situated at the intersection between the tourism market, public action and popular education associations. Suggesting an alternative to commercial tourism, it is represented by the National Union of Tourism and Outdoor Associations (UNAT). This sector is committed to the right to holidays for all, with the intention of promoting access to culture through social support for the public. However, it is possible to hypothesise that there is no specificity in social tourism from the point of view of working conditions.

Keywords Learning · Public action · Popular education associations · Working conditions · Holiday rights · Social tourism · Popular holidays

The social origin of categories is undeniable: "they translate states of the collectivity, first and foremost"(Durkheim 1995). This chapter considers the category of social tourism, whilst remaining mindful that in sociology, "the category par excellence" is totality, society, "the supreme class that contains all other classes"(Durkheim, 1995). Since the late nineteenth century, categories such as holidays, tourism, and to a lesser extent paid holidays, popular education and holiday camps have played a significant social role. The lesser known sub-category of social tourism emerged in the aftermath of World War II and peaked in the early 1970s. It has declined since. The non-profit group *Tourisme et travail* [Tourism and Labour], which operated at the intersection of the trade union movement and of the tourism market, was eventually swept away after an entrepreneurial turn[1]; around the same time, in 1981, the Ministry of Free Time, led by André Henry (a schoolteacher and trade unionist),

---

[1] Devenu ANCAV-TT. (Pattieu, 2009).

F. Lebon (✉)
Université Paris Cité and Université Sorbonne Nouvelle, Paris, France

C. Guibert, B. Réau (eds.), *Employment and Tourism*, SpringerBriefs in Sociology, https://doi.org/10.1007/978-3-031-31659-3_6

briefly placed tourism under the same umbrella as youth and sports. Perhaps significantly, the fifth and final edition of the *Que sais-je?*[2] book on social tourism was published in 1995 (Lanquar & Raynouard, 1995). Then, in 1998, came a new *Que sais-je?* book on cultural tourism, intended to "broaden horizons, seek out knowledge and emotions through the discovery of a heritage and its territory"(Du Cluzeau, 2005). Cultural tourism is regularly described as the "good", distinguished tourism, offering an alternative to mass tourism (Cousin, 2008). Yet, social tourism, which is more or less explicitly geared towards an underprivileged audience, is not dead.

Generally evoked in connection with holidays, the concept of social tourism now draws on the "social economy" to assert a connection between the non-profit group sector – groups generally called associations in France -, specifically in "popular education" (meaning "youth and sports"), and the tourism sector, although relations between the two have traditionally been contrasted and controversial. Popular education, which works with associations, is intended to promote a sense of citizenship and to promote access to knowledge and the right to holidays. On the other hand, tourism refers to (market-based) activities of the transport, hospitality and food industry that relate to travel for business or pleasure (under the definition used by the World Tourism Organization). Neither tourism nor popular education or even the social economy form an easily identifiable group of professions or sector.[3] How, then, can the concepts of "social" and "tourism" go hand in hand?

"The philosophy of social tourism is that every individual is entitled to holidays and that this right should be turned into a reality", according to Jean Froidure (born 1928), a professor of German literature and civilization who founded a research team on tourism in 1985 (Froidure, 2002).[4] In this contribution, which draws on an exploratory research conducted in 2019,[5] the space of "social tourism" is understood as the intersection between tourism and the world of popular education associations. It is characterized by an activism in favour of broadening access to "holidays", meaning "journeys" for non-professional reasons including at least four consecutive nights away from home (under the definition of the World Tourism Organization). Due to its current reference to the social economy, it bases its legitimacy on the status of holiday organizers. For an association, a works council or a local authority, it is possible to claim to do "popular education" or "social tourism". This pertains mainly to the organization of holiday camps and holiday retreats. The "social" label may refer to the price, the audience or even the content of such holi-

---

[2] Que. sais-je? is a popular collection of short books providing accessible introductions to a variety of subjects, similar to the Very Short Introductions in English.

[3] Only "passenger transport" and "tourism" are featured in the occupational nomenclature of the French national institute of statistics.

[4] Froidure Jean, « Du tourisme social à une politique sociale du tourisme », *Informations sociales*, n° 100, 2002, p. 64. His research team on social tourism in Europe was created in 1985 within the Department of Applied Foreign Languages at the University of Toulouse-Le Mirail.

[5] This research consisted in three interviews with national executives from the organizations UNAT, ANCV and APF Évasion), the observation of a study day held by a paramunicipal association (in Gennevilliers) and two interviews with counsellors in a holiday resort.

days. What are the voices, the organizations and the workers involved in these forms of community holidaying? The concept of social tourism in fact relates to the experience of lower-class groups and to the place allocated to state intervention in matters of holidays and leisure, as in other public policy areas such as social housing. Thus the term "social tourism" ultimately refers to a shifting, uncertain reality, a necessary fiction, a category of public policy and activism.

6.1 At the Crossroads of Tourism and Holidays for the Underprivileged

The concept of social tourism comes from a long history of organizing in associations at the intersection of mobilizations around (car-based and bourgeois) "tourism" and (lower-class) "holidays".[6] The use of similar phrases such as "popular tourism" (*populaire* being in French another word for "lower-class" or "working-class") or "mass tourism" would have to be studied. The idea, at any rate, is the same: whereas tourism began as an aristocratic practice, the extension of holidays has broadened its social base. As a result there is a space of social tourism that rests on private and public organizations and is situated at the intersection of the tourism market and of public and activist action (associations, public services, trade unions).

6.1.1 The Promotion of the Sub-Category in the 1960s

The history of many organizations ties in with "social tourism", of which the National union of tourism and outdoors associations (UNAT) became the mouthpiece in the 1960s. The Centre for the archives of youth and popular education associations (Pajep) lists associations whose archives are now often scattered, with names such as Leisure Holidays Tourism (LVT), Tourism in rural areas (TER), Holidays tourism families (VTF), the French federation of popular tourism (FFTP), etc., not to mention a great many associations that organized holiday camps or managed family holiday accommodation centres. Many have disappeared, but some have stood the test of time and still exist today, such as Arts and Life, which was created in 1955 by the National Education Federation (FEN).[7]

Tourism and holidays became increasingly accessible practices in the 1960s, a period characterized by more paid holidays and more automobiles, increases in purchasing power and policies promoting holidays: "the conditions for the economic

---

[6] A new urban bourgeoisie helped broaden the space of tourism, including the swimmers at the Racing club de France who founded holiday clubs in the late 1940s and the "artist hikers" who created the specialized tour operator Terres d'aventure in the 1970s (Réau, 2011).

[7] The role of various trade union confederations should be considered. On CFDT, FO and the CGT (Tartakowsky & Tétard, 2006).

development of a touristic sector" and the "supervision of the free time of the middle and lower classes" were met. Although it had a "subordinate position", social tourism benefited from that dynamic as it earmarked affordable accommodations for specific categories of holidaymakers (Réau, 2011).

Beginning in the 1950s already, the rise of tourism and holidays among the lower classes led some intellectuals to promote the concept of "social tourism". According to the Swiss professor Walter Hunziker, "the twentieth century can be considered as the *era of social tourism*. Social tourism has become part and parcel of the *lives* of ever growing populations; for *masses of people*, it has become a given, in some way a *reason to live*." Social tourism "chiefly relates to social policy and touristic policy". It "tends towards the emancipation of the economically weak classes", allowing them "access" and "participation" in tourism. Its goal "reflects that of tourism in short": regeneration, the need for religion, education, bettering oneself, curing diseases (Hunziker, 1951).

In a 1956 editorial, Paul Mistral, a Socialist Party senator of the Isère region, defined social tourism as follows: "Social tourism is generally understood as the tourism of salaried workers. Actuallly, a genuine social tourism would have society either fully cover the costs of holidays, or to a significant extent financially or organically facilitate trips for some categories of people (Comités d'entreprises, 1956)".[8] However, "social tourism" remains a neglected category, much like "popular education": "If popular education can be described as a quasi-category, it is primarily in the sense of a crypto-category, meaning a little-known label whose existence its promoters must keep proclaiming, concealed as it is by other designations (Chateigner, 2012)."

One organization, UNAT, has emerged as the mouthpiece for the space of social tourism.[9] Since the 1950s, it has monitored measures promoting popular tourism, in a spirit of openness and supervision of the masses by the elites (Lebon, 2020).

6.1.2   UNAT, "the head of the network of social and solidary tourism"

During the 1980s, social tourism went through changes and a critical situation that called for a "renewal" and "reasons to hope". Some authors then associated social tourism with the social economy. The use of the "social economy" category to refer to associations, cooperatives and mutual society is thought to date back to the late 1970s (Rodet, 2019). In his 1986 book on social tourism, Yves Raynouard mentions the "social economy", the "third way", "the "third sector" and its "participationist"

---

[8] Articles in this thematic issue pertained chiefly to tourism (16 pages) and holiday resorts (4 pages); a directory of "popular tourism associations" (6 pages) was also included.

[9] Similarly, in the area of housing, the Union sociale pour l'habitat USH https://www.union-habitat.org/

principle (Raynouard, 1986)..[10] Jacques Chauvin, then a salaried member of the League for teaching and the vice-president of UNAT, also associated social tourism with the social economy in his 2002 book whose title translates as "Social and associative tourism in France: A key actor of the social economy"(Le tourisme social et solidaire: des vacances pour tous et pour tous les goûts, et si tout le monde avait droit aux vacances? un enjeu pour le développement des territoires, 2014).This was a particularly easy connection to make considering that as in the case of youth and popular education associations, the certification of social and familial tourism was founded on adherence to the community-based ideology of the social economy, with a democratic operation, disinterested management, social mixing, activities, etc. The organizations involved are youth hostels, campsites, holiday camps, holiday resorts, educational institutions (for school trips), sport centres (such as UCPA or Glénans), international accommodation centres, etc.

François Soulage (born in 1943), a relation of the former socialist Prime Minister Michel Rocard and one of the architects of the resurgence of the social economy in France, presided UNAT between 1999 and 2008. The preponderance of the reference to the social economy appears to date back to that period. The UNAT now calls itself the "head of the network of social and solidary tourism", which is arguably debatable if one considers product, intended audience, prices, absence of publicity and of for-profit activities. At any rate, its roots in the associative, non-profit world of social economy are openly asserted. According to UNAT's Director-General: "All the actors we are still representing today are active in the field of the social and solidary economy. So that's what UNAT is today, it is the federation, the union that represents, defends the interests of all the tourism operators who are in the non-profit field, with a disinterested management, because their economic model is not connected to private stakeholders. Even though they are market actors, as in they offer... they offer market products, right. But their ultimate goal isn't profit, as they are operating within the framework of the social and solidary economy; their ultimate goal is for people to be able to go on holidays, to offer accessible, quality holidays.".

UNAT brings together large youth and popular education associations that organize, among other things, holiday camps and school trips: the League of teaching, the Léo Lagrange Federation, the United federation of youth hostels, the Union of outdoor sport centres (UCPA), the French union of holiday and leisure centres (UFCV), etc. It also has members among associations in the social sector (APF,,[11] the Little Brothers) and associations or the works councils of public or formerly public companies: La Poste, SNCF (ATC[12]), Banque de France, Air France, Générale des eaux, Ministry of Finances, etc. Last but not least, it includes associations whose main activity specifically pertains to family tourism and "solidary" travel operators: Azureva, Cap France (an offshoot of the Federation of family holiday accommoda-

---

[10] Raynouard Yves, *Le tourisme social. De l'illusion au renouveau?*, Paris, Syros, 1986, p. 11–14.

[11] Association of French paralytics.

[12] Touristic association of railway workers.

tion centres), Cap'vacances, Ternélia, Vacances bleues, Villages club du soleil, VVF, etc.

Michelle Demessine (born 1947) is UNAT's current president. She was a (Communist party) secretary of state for tourism in the Jospin cabinet between 1997 and 2001. There are reportedly around 2000 social tourism facilities in the country (1400 of which are federated under the banner of UNAT), half of which are *villages vacances* (holiday resorts). The sector employed 40,000 workers and hosted four million holidaymakers in 2013 (Atout France, 2016).

6.1.3   Promoting an "Other" Kind of Tourism and the Right to Holidays

The term "social tourism" is a versatile one. In late 2014, members of the European right (e.g., David Cameron) used it to refer to migrants from the poorest EU countries thought to move to other EU countries allegedly to enjoy a more generous welfare state.[13] The same subject has been addressed in France by pundits ranging from Julien Damon (Europe 1) to Éric Zemmour (RTL) and numerous politicians. The term as it is used in this chapter is largely unknown, competing as it is with other meanings.

Since the mid 1990s, the concept of "sustainable tourism" (as well as "cultural tourism") has become increasingly popular. The World Tourism Organization defines it as "tourism that takes full account of its current and future economic, social and environmental impacts". This is a possible niche for UNAT, which promotes solidarity with host populations. "We also try to work on sustainable tourism, because actually many UNAT members have been doing sustainable tourism the whole time without... They just don't necessarily say it out loud, because it's in their DNA as associations, just to... They don't make it a selling point. So the question is how we can get them to highlight this a little better and differently now" (UNAT DG).

"Social and solidary tourism" is a political label that suggests an alternative, another kind of tourism. The sector defends the right to holidays for all, arguing that "the right to holidays and the access to culture are closely connected".[14] According to Luc Greffier, two ideologies coexist: on the one hand, an educational social tourism, which is part of a social policy, and emphasizes socialization through the work of counsellors and on the other, a social tourism that merely broadens access to holidays by lowering prices (and promoting the self-reliance and autonomy of holidaymakers) (Greffier, 2010).

Actors in the social tourism sector are particularly prone to discussing inequality regarding access to holidays. Members of the upper social classes go on holidays

---

[13] *Le Monde*, 14 November 2014.
[14] UNAT general assembly, May 2019.

twice as much as blue-collar workers (Hoibian & Müller, 2015). Press campaigns are regularly staged to assert that the right to holidays of children and youths must be dissociated from market interests, and fostered through accredited associations and local authorities with an ad hoc regulation system. For instance, the Jeunesse au Plein Air (Outdoor Youth) association organizes a solidarity campaign on a yearly basis in educational institutions, under the aegis of the Ministry of Education. On 21 June 2019, it also held a day on the right to holidays at the National Assembly – an event entitled "Access to holidays for all, a social emergency". "Tourism" is often singled out as a market-oriented foil in the world of counselling and popular education.

The director of APF Évasion emphasizes the social implications of the right to holidays for all, and more broadly advocates for an inclusive society: "I'm not in competition mode, since clearly, we leave 500 people hanging every year. As far as I'm concerned, ideally, APF's idea is that every person with a disability can go on a holiday if they want to do so, that's the ultimate goal". She also wants to "push the boundaries" and work towards an inclusive society with regard to employment, training, access to leisure activities and holidays.

6.2 Social and Educational Intentions

From an institutional standpoint, the International Bureau of Social Tourism's Montreal declaration of 1996 symbolically opened up the social tourism sector to the market, noting that "tourism operators cannot look for justification to statutes or procedures, but rather to their actions in pursuit of a clearly stated objective" (Article 15). Thus, "any tourist organization […] which, by its articles of association or statement of aims clearly identifies with social objectives and the aim of making travel and tourism accessible to the greatest number, - thereby differentiating itself from the sole aim of profit maximization - may claim membership of the social tourism movement." (Article 13). Atout France, the French state operator that classifies organizations using star ratings, does not consider the "social" character of some operators.

Social tourism operators rely on private service providers. The association UCPA acquired the Telligo company in 2017. Private, for-profit operators have invested in holiday camps and school trips. Lastly, while the promoters of social tourism highlight political and cultural ambitions, holidaymakers' practices reflect a preference for entertainment and the company of working-class peers.

The distinction between "tourism" and "social tourism" is thus uncertain, considering the evident connections between the two. Tourism dominates in "social tourism". Classifications are "systems of hierarchized notions. Things are not simply arranged by them in the form of isolated groups, but these groups stand in fixed relationships to each other and together form a single whole" (Durkheim & Mauss, 2009). Using a diacritical perception of the social world, "social tourism" cannot be considered without its complement, (economic) tourism, seen as profit-oriented and undemocratic. Such a perception introduces discontinuity where discontinuity is found today between the "left hand" and the "right hand"(Bourdieu, 1998) of tourism.

Despite all these objective limitations, the discourse of promoters of social tourism highlights the existence of a distinct legacy, the ideal of social support, the presence of counsellors, the effort to do things "for and with people". Holidays are something to be learned: this requires understanding and supporting some holidaymakers.

6.2.1 Social Policies

Nearly one in four French citizens received financial aid to go on holidays in 2014: 14 per cent in the form of holiday vouchers, 10 per cent from their employer or works council, 5 per cent from the family allowance office CAF, 1 per cent from their municipality, and 1 per cent from another organization. Overall, however, members of the most privileged groups benefit more from this aid (Hoibian & Müller, 2015). There are different sources of funding of social tourism: the state, CAF, local authorities and organizations with social missions, such as works councils, the social services of public companies, pension funds or mutual societies.

Social tourism is backed by social policies. Holiday vouchers (*chèques-vacances*), which were introduced in 1982, are managed by a dedicated national agency, ANCV. They are distributed through works councils and among public sector employees to a total of roughly six million persons. However, these vouchers are massively spent on services provided by for-profit operators, which tends to "open up" or "water down"(Greffier, 2006) whatever specificities social tourism may exhibit. Still, the ANCV's operating surpluses are earmarked for policies aimed at aiding access to holidays for the most underprivileged fractions, those left out of salaried work. This includes the two schemes "Seniors on holidays" and "Holidays 18–25" (Source: ANCV social policies officer).

The goal here is to facilitate financial access to holidays by lowering prices. As the APF Évasion manager explains: "It works thanks to a wide array of aids that can be granted. There are aids at the level of the MDPH (departmental houses for the disabled), which contribute up to a certain amount. Through the ANCV, which is a big partner, and creates a big fund using the companies' unused holiday vouchers, etc. This fund goes out to disabled people. It's huge, we're talking millions. With the ANCV, we examine holidaymakers' applications and sometimes we'll cover 800/900 €, divide the cost in two. Plus it's cumulative; each holidaymaker can apply for financial aids, or else they might have the means to go without doing this".

In addition to holiday vouchers, there are also vouchers offered by the family allowance office (CAF). CAF's holiday aid scheme[15] contributes to the

---

[15] https://www.vacaf.org/

reconstruction of self-images, the creation of happy family memories and the reintroduction of recipients into local sociability networks (Guillaudeux & Philip, 2013; Guillaudeux & Philip, 2014). Some associations have openly social goals. For instance, *Vacances et familles* (Holidays and families), which relies on a "network of committed volunteers and employees" assists families before, during and after their stay. Likewise, the action of *Vacances ouvertes* (Open holidays), which was founded in 1990 by Edmond Maire (1931–2017), fits within a broader commitment to combating all forms of inequality and exclusion.

The space of social tourism tends to be split between more or less socially-minded actors, especially considering it targets a precarious audience. There is a tension between the "social" and the "cultural", meaning the working-class and the bourgeois, the lowbrow and the highbrow (some departmental tourism committees are eager to go "upscale", which involves for instance limiting the number of campsites). Jean Froidure argues that there is a "fracture" in social tourism, with "on the one hand, a traditional social tourism inherited from the Trente Glorieuses [the country's post-World War Two boom years], which is almost exclusively geared towards some workers; on the other, we have a new social tourism that is about social work, and that essentially targets the new populations who are excluded from employment or job stability"(Froidure, 2002).

Holidays are sometimes envisioned as opportunities for social work (which may involve budgeting, fostering ties between parents and children, etc.), which raises the question of the publics of social tourism. The norm of holidaying is unequally distributed. The absence of holiday trips is experienced differently among working-class families, from a feeling of exclusion (which is most common) to indifference that may even border on lack of awareness of the possibility of such a thing as travelling for the holidays, replaced for instance by domestic work (Périer, 2000).

6.2.2   Learning to Travel

The social sciences, and especially the learning sciences, have investigated tourism, as one needs to learn to travel and sometimes to travel to learn, which may happen in a school environment (in this sense, school trips are social tourism). The practice of tourism requires skills that some do not have: choosing a destination, booking a ticket, an accommodation, packing, budgeting, finding one's way around, living and in some cases collaborating with others, undressing, sunbathing, dressing in an unfamiliar setting, making oneself understood in a foreign language, speaking about oneself, etc.

Travelling for tourism thus involves acquiring skills encouraged by social workers and "serves a social integration project for recipients, allowing them more generally to gain autonomy and self-esteem. In this way, tourism can be deeply emancipatory, and has precisely been presented as such by trade union organizations and more broadly activist education movements"(Peyvel & Löfgren, 2019).

The project of developing an educational tourism has been studied by researchers such as Luc Greffier and education sciences professor Gilles Brougère. In a 2012 article, Brougère built on the theories on "situated learning" within the framework of a research on social tourism (a group of women and their children supported by the social centre of a city in Northern France). He proposed the concepts of "guided participation" and "guided exploration" to address the learning effects of tourism practices (Brougère, 2012). He has edited books on learning in a touristic setting (Brougère, 2012)[16] and on school field trips (Colin, 2009; Brougère & Wulf, 2018).

Professionals and holidaymakers even learn from one another, particularly when professionals/activists have precarious life trajectories that resemble those of holidaymakers (Ulmann, 2014a; Ulmann, 2014b). However, the influx of customers from the pauperized and racialized lower classes, devoid of "touristic capital", is sometimes experienced as a regrettable form of "ghettoization" by professionals who promote education, secularism and social mixing (Fabbiano, 2014; Fabbiano, 2015).

6.2.3 Working in (Social) Tourism

The concept of "social tourism" is unknown to the public at large and even to most of the workers whose work can be said to fall under that category. They may describe themselves as working in counselling, cooking, etc., sometimes with a reference to the tourism sector, but they do not know what "social tourism" is.

Managers stress the non-profit status of most employers, while acknowledging that they are operating in a professional world characterized by the pursuit of profit: "a VVF holiday resort is a non-profit association. So our customers are essentially families who come during the school holidays; we also have family groups or just individuals. It's a tourism organization that is open to all. A non-profit one, I will insist on that. You know, we're not in it to make money really, we're in it to be accessible, to be close, to be convivial with our customers".

Work in "social tourism" and by extension in the non-profit sector encompasses a variety of statuses from volunteering to salaried work. The highly variable share of volunteer work in organizations is probably higher in social and popular education associations, which have for instance seen mobilizations in the defence of statuses that are effectively derogations to the French Labour Code. The tourism sector appears to be a laboratory for broader transformations of labour. Social tourism especially raises the question of the place of volunteering in organizations.

For instance, the manager of APF Évasion, mentions challenges related to the recruitment of volunteers: "In my view, our model has overall been fragilized in a

---

[16] The first part of the book coordinated by Gilles Brougère and Giulia Fabbiano, op. cit. is entitled "Social tourism and educational projects".

societal sense". She contends that our consumer society is a hindrance to collective engagement: "We consume, we consume products, but making the effort to contribute, to tell yourself, 'I'm part of an association, and I'm going to go ahead and do things', it's less frequent among young people today". The association seeks to hire driven volunteers rather than simply young people trying to have a nice holiday at a popular destination. In the same register, calls for willing individuals are frequent, particularly in the local press: "To ensure that 700 persons with disabilities enjoy a vacation, the association *Un Brin de soleil* hires 200 counsellors each summer". These holidaymakers, which work in ESATs (specialized work-based support establishments), are monitored by counellors under educational engagement contracts, paid a gross sum of 48€ a day (Ouest France, 2019).Volunteering dominates, and salaried work has historically been and still more scarce (Simonet & Lebon, 2012).

Yet, in addition to the volunteers, there are also salaried "social tourism" workers, exhibiting a wide variety of statuses, from seasonal student workers employed during school holidays to the cooks, cleaners, etc. who attempt to work all year, sometimes juggling multiple jobs.

Here, I focus on the counselling sector, which appears to have a central role in the management of work organizations, holiday camps and holiday resorts. Counsellors also make up the bulk of the seasonal workforce, and three in four social tourism jobs are seasonal.

Evidently, trajectories leading to counselling in social tourism follow paths that run largely outside of the fields of counselling and social tourism. This is unsurprising considering that early vocations are seldom observed in counsellors. In a southern France VVF resort, for instance, the director is a former accountant. The 26-year-old manager in charge of counselling has worked in the retail sector (with the BHV chain of shops) and for the Marmara travel agency. "I went on a holiday, at the time I was with two girlfriends, and I got to see this counsellor job and I really liked it, so I became interested, I asked questions when I got back home, I applied with a variety of tour operators […] I worked with Marmara for five years, so I discovered this job thanks to my holidays, I worked with Marmara and then three years later I became an counselling manager and then I wanted to move on to something else, and that's why I'm here at VVF". Such trajectories tend to support the thesis of an absence of specificity of social tourism from the workers' point of view. The 44-year-old director switched to counselling during his studies: "I was trained as an accountant and I wasn't necessarily happy in that field; during my training, I got into working summers in holiday resorts to pay for some of my studies as a counsellor, and then I met people and became a year-round counsellor, I became a counselling manager for big companies. In 2006, I left counselling to become a director." While the education system socializes individuals and helps place them on the labour market, there is no strict match between training programmes and jobs in (social) tourism. The supply of training for tourism jobs is abundant and varied, reflecting a scattered field of activities (comprising over thirty different branches) (Michun & Guitton, 2006).

In their holiday resort, staffing needs vary widely – there are three salaried staff during the winter and over 50 at "peak seasonal activity", according to the director.

This results in a mass recourse to short-term contracts, including for middle management positions: "I'm sort of a seasonal worker, I'll live in a given place for six months and in another for six months, and with a little bit of time at home, so let's say, 40 percent in a club or a resort, 20 percent at home and 40 percent in a different club", says the counselling manager. Seasonality is sometimes viewed negatively, especially by the parents. Such job-related challenges are sometimes compensated by a sense that one has turned a "passion" (for sport, children, etc.) into a job. In such cases, work and lifestyle blend together (see Christophe Guibert's chapter in this volume).

Employers and managers seek out counsellors who display a sort of joyful flexibility. "Their specificities are conviviality and availability", the director says. "You don't really keep track of the hours", and effectively end up working from around 9 am to 11 pm. Customers are believed to seek out friendly interactions with staff, but also to expect dynamism, a smiling demeanour, etc. These supposedly rewarding working relations have ambiguous effects on salaried workers in tourism. "Each of the workers of enrichment is forced to become their own exploiter as a seller of themselves, meaning that they are both the merchant and the commodity. In this regime, the indefinite extension of individual working time is decoupled from the gains obtained by each worker"(Boltanski & Esquerre, 2020).

In service relationships, and therefore in tourism, keeping the right distance is crucial. Ways of doing this probably vary quite significantly. The VVF counselling manager interviewed here sometimes cultivates relationships long after the work interaction has ended: "People come to see, they cheer for us and they remember our names, I also see it through Facebook, we meet a lot of people and they get attached to us. It can be a little bit tough because we see over 300 people on a weekly basis, but as I was saying the holidaymakers come once a year and they only remember the counselling team they met, so I'm still in touch with holidaymakers I met three to four years ago. Sometimes we chat with people, we tell them about our lives and they tell us about theirs'".

Improving employment and working conditions has been an ongoing concern for the tourism sector and for public authorities. However, it is not certain that the conditions experienced by social tourism workers are better than in tourism. Precarity is the rule everywhere, even in works councils managed by the CGT trade union (Dethyre, 2007; Lefèvre, 2013).[17]

---

[17]Dethyre Richard, *Avec les saisonniers. Une expérience de transformation du travail dans le tourisme social*, Paris, La Dispute, 2007. Sociologue et militant associatif, il a enquêté dans le plus gros CE de France, la Caisse centrale d'activité sociales (CCAS), dirigée par la CGT. Cf. aussi Lefèvre Denis, *Voyage au cœur de la CCAS*, Paris, Cherche midi, 2013.

6.3 Conclusion

Categories contribute to stabilizing social life. To a certain extent, they may even create the realities to which they are applied. Social tourism might seem like a trifling matter for research. The practice of writing, which consists, among other things, in "drawing up lists", classifies and hierarchizes (Goody et al., 1979). Tourism and the holidays are the central, dominant categories here, sometimes inflected by qualifiers such as social, solidary, sustainable, popular, etc.

The promoters of social tourism are trying to fill the gap between the right to holidays and reality: "Those excluded from holidaying and recreational activity are still many, a great many, too many"(Froidure, 1992). Social tourism, which calls for opening up leisure activities and holidays to the broadest possible range of people, is subject to investments by public policies, state authorities and a variety of non-profit supporters.

It would probably be easy to show that social tourism, much like the social economy, does not actually exist in the way that it is depicted, displayed by its most ardent backers. Still, the concept does raise the question of the social role of tourism. Who should ensure that this role is played: the associations, the state, the market, and at which scale? Do we need a social economy of touristic salvation or salvation through the market-based economy? These questions have also been raised in other fields, such as education or housing. The concept of social tourism matches a tourism policy, and is a way among others of questioning the social, economic and cultural implications of tourism and holidays in social relationships.

References

Atout France. (2016). *Les valeurs ajoutées du tourisme social et solidaire.* Editions Atout France.
Boltanski, L., & Esquerre, A. (2020). *Enrichment: A critique of commodities (Poter, C. Tran.).* Polity Press.
Bourdieu, P. (1998). *Contre-feux I* (p. 9). Liber-Raisons d'Agir.
Brougère, G. (2012). Pratiques touristiques et apprentissages. *Mondes du Tourisme, 5,* 62–75.
Brougère, G., & Wulf, C. (2018). *À la rencontre de l'autre: lieux, corps, sens dans les échanges scolaires.* Téraèdre.
Chateigner, F. (2012). *" Education populaire": les deux ou trois vies d'une formule* (Doctoral dissertation, Strasbourg). 20.
Colin, L. (2009). Chapitre 6. Les séjours à l'étranger : apprendre malgré l'institution scolaire ? In G. Brougère (Ed.), *Apprendre de la vie quotidienne* (pp. 21–31). Presses universitaires de France.
Comités d'entreprises. (1956). Liaisons sociales. *Tourisme Social, 347,* 3.
Cousin, S. (2008). L'Unesco et la doctrine du tourisme culturel. Généalogie d'un «bon» tourisme. *Civilisations. Revue Internationale d'Anthropologie et de Sciences Humaines, 57,* 41–56.
Dethyre, R. (2007). Avec les saisonniers. In *Une expérience de transformation du travail dans le tourisme social.* La Dispute.
Du Cluzeau, C. O. (2005). *Le tourisme culturel* (p. 3). PUF.

Durkheim, É. (1995). *The elementary forms of religious life.* (Karen E. Fields trans. Free Press.

Durkheim, E., & Mauss, M. (2009). *Primitive classification.* (Needham, R. trans (p. 48). Routledge.

Fabbiano, G. (2014). "Ces familles-là ne savent pas ce que c'est les vacances", discours et représentations du tourisme social. In G. Brougère & G. Fabbiano (Eds.), *Apprentissages en situation touristique* (pp. 33–53). Villeneuve d'Ascq, Presses universitaires du Septentrion.

Fabbiano, G. (2015). Rhétoriques d'altérité et représentations de l'(im) mobilité: les dynamiques d'ethnicisation à l'œuvre dans le tourisme social. *Sociologies.*

Froidure, J. (1992). Pourquoi des cahiers? In *Tourisme social en Europe* (Vol. 1, p. 7). Université de Toulouse Le Mirail.

Froidure, J. (2002). Du tourisme social à une politique sociale du tourisme. *Informations Sociales, 100,* 64–73.

Goody, J., Bazin, J., & Bensa, A. (1979). *La raison graphique: la domestication de la pensée sauvage* (p. 187). Editions de Minuit.

Greffier, L. (2006). *L'animation des territoires: Les villages de vacances du tourisme social* (pp. 65–69). L'animation des territoires. Paris: L'Harmattan.

Greffier, L. (2010). Le Tourisme social et associatif: illusion entretenue ou contexte singulier? *Sud-Ouest Européen. Revue Géographique des Pyrénées et du Sud-Ouest, 29,* 65–76.

Guillaudeux, V., & Philip, F. (2013). Etude sur l'accompagnement au départ en vacances familiales. *Dossiers D'Etudes-Cnaf, 162.*

Guillaudeux, V., & Philip, F. (2014). L'accompagnement social au départ en vacances. *Informations Sociales, 1,* 101–108.

Hoibian, S., & Müller, J. (2015). Vacances 2014: L'éclaircie. *CREDOC, Collection des Rapports, 320.*

Hunziker, W. (1951). Le tourisme social. In *Caractères et problèmes* (pp. 7–17). Alliance internationale de tourisme.

Lanquar, R., & Raynouard, Y. (1995). *Le tourisme social et associatif.* PUF.

Le tourisme social et solidaire : des vacances pour tous et pour tous les goûts, et si tout le monde avait droit aux vacances? un enjeu pour le développement des territoires (2014). Paris: Alternatives Economiques.

Lebon, F. (2020). L'UNAT. De la circulation automobile aux vacances pour tous (1920-1974). *Juristourisme, 230.*

Lefèvre, D. (2013). *Voyage au cœur de la CCAS.* Cherche Midi.

Michun, S., & Guitton, C. (2006). Les métiers et formations du tourisme. Logiques des branches professionnelles et perspectives régionales. *Bref Cereq, 233.*

Ouest France (2019).L'association Un Brin de soleil recrute des animateurs pour les vacances. *Ouest France.* https://www.ouest-france.fr/pays-de-la-loire/aigne-72650/aigne-l-association-un-brin-de-soleil-recrute-des-animateurs-pour-les-vacances-6337804

Pattieu, S. (2009). *Tourisme et travail: de l'éducation populaire au secteur marchand (1945–1985).* Presses de Sciences Po.

Périer, P. (2000). *Vacances populaires: images, pratiques et mémoire.* PUR.

Peyvel, E., & Löfgren, O. (2019). *L'éducation au voyage: pratiques touristiques et circulations des savoirs* (pp. 25–26). Rennes.

Raynouard, Y. (1986). *Le tourisme social: de l'illusion au renouveau?* (pp. 11–14). Syros.

Réau, B. (2011). *Les Français et les vacances. Sociologie des pratiques et offres de loisir.* CNRS éditions.

Rodet, D. (2019). L'économie sociale et solidaire: une réalité composite issue d'histoires plurielles. *Informations Sociales, 199*(1), 14–25.

Simonet, M., & Lebon, F. (2012). Le travail en'colos': Le salariat en vacance? *Notes de l'Institut européen du salariat, 26.*

Tartakowsky, D., & Tétard, F. (2006). *Syndicats et associations: Concurrence ou complémentarité?* Rennes.

Ulmann, A.-L. (2014a). Les professionnels du tourisme social : des modes d'agir contre la barrière culturelle. In G. Brougère & G. Fabbiano (Eds.), *Apprentissages en situation touristique* (pp. 55–70). Villeneuve d'Ascq, Presses Universitaires du Septentrion.

Ulmann, A.-L. (2014b). Vacances, service (social) compris. Divertir et militer comme pratique éducative. *Informations Sociales, 181,* 114–122.

Francis Lebon is a professor in education sciences at the Université Paris Cité, member of the CERLIS, CNRS (UMR 8070). His research focuses on popular education professionals and the division of labour in primary schools. He is currently conducting two surveys: one on the policies and practices of access to law; the other on teaching practices in the first grade.

Chapter 7
The New Configurations of Labour in the Tourism Sector: Is Entrepreneurship a Choice?

Aurélie Condevaux and Sébastien Jacquot

Abstract The economy of tourism has been diversifying along with the rise of platforms which offer a variety of accessible, combinable and customizable services to visitors, regarding both accommodations and tours, drawing on evaluations as a tool for recommendation and exposure. Even though this is not systematic, this platform economy is often associated with the rise of the self-employed micro-entrepreneur status. Our research focuses on the ways in which the tourism sector is impacted by these transformations and these new forms of work and of relationships to work.

More specifically, we examine three main aspects relating to the spread of entrepreneurship for small-scale activities often performed through platforms: first, how workers experience new forms of labour, from subordination to more or less deliberately chosen forms of limited autonomy and claims of freedom. Then, we consider the know-how and skills used by these workers, the capitals they mobilize, and the effort these new forms of labour require in terms of managing one's status and organizational (re)positioning. Behind the veil of a language that negates the ideas of labour and constraint, we also need to evidence the schemes used to hire and evaluate these workers. Lastly, we investigate the intersection between work and non-work, in order to "identify the new frontiers of the employment relationship" (D'Amours et al Relations industrielles, 72, 409–432, 2017).

Keywords Tourism · Work · Entrepreneurship · Digital labor · Plateforms

The economy of tourism has been diversifying along with the rise of platforms that bypass some of the traditional actors (destination management companies, public and para-public bodies specialized in promoting tourism and catering to tourists). These platforms offer a variety of accessible, combinable and customizable services to visitors, regarding both accommodations and tours, drawing on evaluations as a

A. Condevaux (✉) · S. Jacquot
IREST (Institute for Research and Higher Studies on Tourism), Paris, France
e-mail: Aurelie.Condevaux@univ-paris1.fr

tool for recommendation and exposure. In terms of labour, these new organizations of tourism lead to an increased plurality of work statuses, many of which tend to fall outside the realm of wage work. The activities offered by these many platforms – Airbnb and Airbnb Experiences (Chareyron & Jacquot, 2017), TripAdvisor, *RendezvousCheznous* – are not limited to guided tours. They combine the basic skillset of tourism with a wide variety of forms of know-how and experiences (from creating perfume to massages, wine and cheese tastings, or a meeting with a professional athlete). Platforms have also emerged in specific tourism-related jobs, such as *Conciergering*, which allows individuals, sometimes salaried in another capacity, to find extra work as self-employed concierges.

These changes, of course, have certainly not only been happening in the tourism economy. In recent years, social science studies on labour have increasingly examined the forms of work induced by platforms, using umbrella terms such as "platform capitalism" (Abdelnour, 2019), "collaborative economy" (Rodet, 2019) or "digital work" (Scholz, 2016). These phrases refer to varied situations, from the appropriation of free leisure activities performed by consumers that generate profit for digital companies (e.g., Facebook's likes) to forms of labour outsourcing, as some of the tasks offered through these platforms can resemble real full-time jobs, when they are open-ended assignments requiring a high number of working hours (Abdelnour, 2019), as in the case of ridesharing drivers. This platform economy is often associated with the rise of the self-employed micro-entrepreneur status. Yet, this is not systematic, and other forms of labour are observed: informal work, volunteering, extra work, and in some cases wage work when individuals are affiliated with a company that offers services through a platform.

Our research focuses on the ways in which the transformations of wage work – and labour in general – impact the tourism sector, leading to new forms of work and of relationships to work. More specifically, we consider the spread of entrepreneurship – both as a legal category and as a form of engagement in labour – for small-scale activities often performed through platforms.

In doing so, we pay attention to three main aspects:

First, we examine how the new forms of labour are incentivized and made imperative, as well as the ways in which workers experience them, from subordination to more or less deliberately chosen forms of limited autonomy and claims of freedom. This requires being mindful of the workers' trajectories and their relationships to the platforms.

Then, we consider the know-how and skills used by these workers, the capitals they mobilize, and the effort these new forms of labour require in terms of managing one's status and organizational (re)positioning, with the rise of the micro-entrepreneur and other self-employed statuses and the recourse to intermediaries such as associations and training institutes. Behind the veil of a language that negates the ideas of labour and constraint, we also need to evidence the schemes used to hire and evaluate these workers: competences and qualifications are now less important than "expertise" and soft skills, which are no longer just assessed by a recruiter, but directly and constantly by clients. In addition, work is also needed to promote one's activity, which requires digital communication and marketing skills.

Lastly, we investigate the intersection between work and non-work, in order to "identify the new frontiers of the employment relationship" (D'Amours et al., 2017). While the existence of "work on the side" has been documented before (Weber, 2009), this kind of work has experienced a renewal with "digital work", which raises questions that are specific to the context of tourism. On the one hand, research on employment in the touristic sector has abundantly documented the strategies used to dissimulate market-based and employment relationships to ensure an enchanted tourist experience (Réau, 2006; Giraud 2007) – indeed, the tourist experience is a particular kind of "commodity", based on social relationships and ephemeral performances (MacCarthy 2007). On the other hand, this again raises the question of the commodification of free time (Abdelnour, 2019) and the balance of personal life and work, as the two spheres are increasingly porous, in terms of space (one's home can for instance be an instrument of work), activities (often connected to a "passion" or amateur practice) and of the transactions they involve.

This contribution draws on multiple sources. First, in our capacity as lecturers and researchers at the University Paris 1 Panthéon Sorbonne's Institute of Research and Higher Studies on Tourism (IREST), we began constructing these research questions after supervising a workshop in the field during the academic years 2017–2018 and 2018–2019. In this workshop, first-year students from the DATT master's program (Development and Territorial Tourism Planning) tackled the subject of tourism employment in Plaine Commune, a public intermunicipal body within the Greater Paris region. This workshop was conceived with the *mission tourisme*, a subdivision of the Directorate for economic development's social and solidary economy department. The goal was to measure the contribution of tourism to local employment, and to study the new forms of tourism work and their impact in terms of employment, including the rise of temporary work, outsourcing, and micro-entrepreneurship in the tourist and hotel industries (IREST, 2018; , 2019). In the process, the students identified self-employment dynamics, particularly in the field of mobility, through the case of ridesharing drivers (a sector that goes beyond tourism) and in the multiform sector of tourist guiding, through the mediation of online marketplaces. In addition to insights gained in the course of supervising this workshop, we draw on interviews conducted by students in 2018–2019 – particularly three interviews with individuals working in entrepreneurship support facilities (ADIE -Association for the Right to Economic Initiative - and Chambers of Commerce and Industry), and interviews with temp workers and self-employed tourism workers.

Building on these initial efforts, we launched interview surveys targeting autonomous tourism workers that are affiliated with a platform, often under the French micro-enterprise regime,[1] with jobs in hospitality, tourist services (in sport or cooking), guiding and influence, in Paris or close surroundings. These individuals did not necessarily plan to work in a touristic setting, but they effectively work

---

[1] This status has various equivalents in European countries: Ich-AG (Germany), trabajadores autonomos (Spain), Parasubordinati (Italy) (Abdelnour, 2017).

with – mainly international – tourists. This research was also extended during the pandemic by one of the authors by interviewing people offering online tours or experiences on tourism platforms in other destinations. We consider this contribution as a first step in a research that is still in its infancy and on which we will further expand.

In the following, we first present existing approaches to entrepreneurial dynamics in tourism studies. Then, we address the questions of entrepreneurship in tourism from three angles that allow us to consider the different facets of our object of study in an exploratory manner: the profiles and situations of the workers, their know-how and skills and how they are monitored by the platforms, and lastly, the permeability of the spheres of work and leisure.

7.1 A Sudden Rise of Entrepreneurial Logics in the New Forms of Tourism Labour?

Understanding the new rationales of labour in the field of tourism, and their borrowings from entrepreneurial models or standards, requires situating these changes in the studies of the links between tourism and labour. Tourism as labour has not been a major concern for tourism studies. In 1978, tourism geographer Robert A. Britton noted the lack of interest in this question, which he attributed to the fact that these jobs were seen as low-skilled and precarious, at a time when wage work still dominated largely (Debbage & Ioannides, 1998). Over thirty years later, Soile Veijola (2010) made a similar observation, pointing to the low number of studies on the links between tourism and employment, which focus mainly on the hotel and lodging industry. A few approaches have emerged in the 2000s and 2010s (Ladkin, 2011; Ioannides & Zampoukos, 2018), from the management approach to a cultural approach (tourism as performance by employees), and an approach focusing on inequalities, particularly regarding gender, through the study of room attendants (Puech, 2004; Guégnard & Mériot, 2010).

Arguably, this difficulty of identifying a field of study on tourism as labour might be the result of the wide variety of jobs and statuses in the sector. The study of employment in tourism is a challenge in terms of delimitation. If we consider that tourism jobs encompass all jobs that involve interacting with tourists, this induces a great thematic variety (from transport to guiding or lodging) and frequent doubts as to the strictly touristic nature of the work. In France, different attempts to delineate the field of tourism jobs have been made in reports. The Céreq report of 2006 identifies five "poles" (Guitton & Labruyère, 2006), whereas the Nogué report of 2013 lists eight "families" of jobs.

This variety is not only thematic; the sector is also characterized by a wide range of statuses. Adopting a regulationist approach applied to tourism, Ioannides and Debbage (Debbage & Ioannides, 1998) connect particular forms of economic organization with distinct forms of labour. The Fordist economy is associated with low-flexibility jobs, requiring the performance of repetitive, low-skilled and badly paid tasks, characterized by frequent turnover and seasonality, whereas post-Fordism leads to a bipolarization of the labour market, with a few highly skilled core jobs on the one hand, and large numbers of generally low-skilled short-term jobs in the form of outsourcing or temp work on the other. These transformations in the forms of labour are not specific to tourism (Castel, 2009), but have specific outcomes: as in the cultural sector, as the increasingly widespread language of creativity sweeps domination and precarity under the rug (Boltanski & Esquerre, 2017), tourism jobs can be at the same time described as life choices, extensions of an ethos of exchange or a passion turned into a vocation, and be analysed in terms of undesired precarity.

Amid this diversification of statuses and forms of labour, entrepreneurial types of jobs and the associated language have been on the rise for a wide range of tourism-related activities.

Entrepreneurship is a statistically fuzzy category, defined by the entrepreneur's "particular logic of action" (Chauvin, 2010). The main features of entrepreneurship that are usually cited are capital raising, independence (in contrast with wage work) or a discourse emphasizing independence, responsibility in the organization and risk-taking in a context defined as uncertain or competitive (Menger, 2014). However, this microsocial approach to entrepreneurship should be complemented with a macrosocial approach, positioning it among its meanings, supporting structures and networks (Chauvin et al. 2014). The term is increasingly popular and ideologically loaded, to the effect for instance that it is presented as a means of returning to work for individuals in precarious situation, as in France with the new, simplified *auto-entrepreneur* self-employed status (Abdelnour, 2013). The spread of the term also reflects a particular mode of subjectification (Dardot & Laval, 2010) that characterizes the entrepreneurial ethos. The entrepreneurship category is now used to described and analyse varied situations, which includes the forms of engagement of individuals who have little capital in tourism work.

Until recently, small-scale firms were rarely studied (Shaw & Williams, 1998), in comparison to large companies in the fields of transport, hospitality and tour operating, due to the influence of an evolutionist approach whereby the intensity of tourism development was related to the size of the locally active operators, following the model of the tourist area life cycle (Butler, 1980). The current changes in the tourism economy have however undermined the relevance of this approach, as multiple activities are externalized (with a growing recourse to outsourcing), an increasingly multifaceted form of labour, encouraged by platform capitalism (Abdelnour, 2019), has developed, often under the legal status of the micro-enterprise, and multiple start-up companies have popped up in the field (Queige, 2015), against the backdrop of a quest for digital and commercial innovations. These new forms of labour have been studied in particular in the transport sector – not specifically emphasizing tourism – as in the case of Uber (Azaïs et al., 2017; Nasom-Tissandier & Sweeney, 2019), or in Airbnb, but rarely in the context of other jobs, such as tourist guiding (Weiler & Black, 2015).

However, the entrepreneurial dimension can be nuanced, when these new activities do not involve a significant amount of capital and those who perform them do not always define themselves as entrepreneurs (Williams and Shaw 1998). They may highlight other professional identities, present their initiative and career change as the reflection of a passion for the practice rather than of a strictly economic rationale (Dubois & Terral, 2014a; b), or an intimate sense of attachment to a place when former tourists set up a more or less profitable tourism business there (Joly, 2020). These activities can also be carried out on a part-time basis, or as a complement to other activities, or informally, which contributes to their invisibilization (Krinsky & Simonet, 2012) and their place in the "grey areas" of employment (Supiot, 2000). What transformations of labour do these new forms of activity associated with platform capitalism reveal?

7.2   A Constrained Choice or an Affirmation of Freedom?

The above raises the question of whether opting for a micro-entrepreneur status is embraced as a genuine choice or experienced as a constraint, imposed by the new structure of the tourism economy, characterized by the rise of intermediary platforms.

Sarah Abdelnour's (Abdelnour, 2017) analysis of France's *auto-entrepreneur* scheme highlights the precarity that dominates among those who use this status, the "constrained freelancers". Among these micro-entrepreneurs, she finds a minority who use their freelancing income as a supplement, whereas working under this status is often a constraint that allows a contractor not to have to salary the worker or is used as an informal probation period. Studies on the platform economy support this: promises of substantial remuneration are often deceptive. Food bloggers and webmasters, for instance, for the most part do not generate enough income to make a living (Jourdain & Naulin, 2019). In some cases, the forms of exploitation generated by these platforms have warranted the use of the neologism "cybertariat" (Abdelnour, 2019).

However, our first interviews show that users of the micro-entrepreneur status and/or of platforms in the tourism sector exhibit profiles that differ from those described by Abdelnour. They have higher levels of educational attainment (bachelor's or master's) or a significant training in the artistic practice around which the experience they sell revolves; some even have multiple degrees and trainings under their belt. Before developing their current projects, which in some cases are part of a career change, our interviewees – aged 30 to 50 – have held varied but generally skilled positions as journalist, schoolteacher, marketing executive or communication consultant. Some have university-level training in areas that have little do with their current project. For example, a young woman was trained in theatre, yoga and reiki after a stint in a business school. The development of their projects results from multiple motivations and factors, such as the desire to find self-fulfilment in a "dream" job, the need to find a new work following a migration (for an American woman who moved to Paris) or a return to one's country of origin after several years spent abroad (for a French woman returning to that same city). In other cases, developing an online activity through plateforms is a response to the need to find new sources of income to make up for the sudden halt of tourism as a result of the Covid-19 crisis.

In several cases, opting for an entrepreneurial status and using a platform are the result of a decision, sometimes made after consulting other people. One of our interviewees decided to offer activities through a platform and with a micro-entrepreneur status as part of a career change that she devised after an initial "successful" experience of entrepreneurship (as the manager of a limited liability company). In the cases we have observed so far, the choice of status is made after the person herself has consulted someone she considers competent (here, the former accountant of her company), or based on the experience of friends, relatives or acquaintances with a similar background. Yet, the choice of status also appears to be made by default, in the context of difficult job search experiences. In two cases, the micro-entrepreneur status is viewed as a first step in a trajectory towards something else (a company operating under a different status, or a start-up). The status is not experienced as a form of wage work in disguise, or the result of external incentives or constraints: none of our interviewees has a "single client" that would actually function as an "employer/client" (per the phrase used by Sarah Abdelnour, 2017 to refer to cases in which the legal "client" behaves effectively like an employer). The status is also sometimes picked for legal needs: for artists in France, the fee-based remuneration system is not suited to touristic services. Conversely, for some interviewees, the work associated with their self-employed status and/or platforms is envisioned or experienced as a form of "freedom". Such a sentiment echoes the promises of independence and emancipation made by digital companies (Jourdain & Naulin, 2019: 49), liberating workers from the employment relationship and the subordination it entails, although mobilizations of workers for firms such as Deliveroo and Uber have precisely denounced the existence of forms of subordination to the platform.

The creation of an activity commercialized through the platform frequently occurs on the advice of a friend, relative or acquaintance. As an interviewee puts it: "I started in 2018, that's when I started working as a yoga teacher and a friend, well, actually a work friend, when I started my yoga work, it's a little bit competitive when you're looking for your first job, she told me 'yeah, I've travelled, I've tried this Airbnb thing, it's new, why not try'. I said, well, okay, I've never used this Airbnb thing, and I had a look, and thought, right, I can get this going, and in 2018 I had my first client". This shows additionally that the platform serves as a distinctive marketing channel in a highly competitive environment. While they also point out that they were encouraged by friends, others first used Airbnb as hosts before they went on to sell "experiences".

No criticisms were voiced by interviewees regarding their relationship to the platform. Some report being very satisfied by the assistance and advantages it brings, despite the costs: "I really think Airbnb is a very good platform for this, because they're connected, and their fee matches what I would have had to spend, say, on Facebook Ads. In the end I'm doing fairly well for the number of classes and the rewarding encounters. Yeah, I do think Airbnb is a very good platform". The platform is also perceived as a means to assess the credibility and trustworthiness of potential clients, especially for those whose clients come to their home: "if I haven't had an exchange with the person first, if they've got zero comments, you know, even a woman, because I see them alone at my place, plus you can't hear a thing, I mean, I do watch out".

However, the status is also associated with major constraints, pertaining especially to patterns of working time. This affects holidays, as in the case of this micro-entrepreneur who offers yoga and reiki classes: "Well, as for holidays, I took a lot of days off this summer, I caught up so it's alright, but then it's tough, I don't really do actual holidays, breaks, life as an *auto-entrepreneur* actually means you don't have your weekends, or your Mondays". The organization of the week can also be problematic, as in the case of a worker on an open-ended contract who also does freelance guided tours: "So [I work] every other weekend at the Basilica as an agent and sometimes as a freelancer if the opportunity arises... You know, we all need money these days, so once in a while, if they tell us to come in on a Sunday, you're not going to bite the hand that feeds. You sacrifice your weekend, which is too bad, but at least... As an agent you clearly have to come in every other weekend and sometimes more, same thing if I have freelance assignments or if there are bank holidays."

Despite these caveats, the overall positive experience of the micro-entrepreneur status and/or the relationship to the platform on display here could be a result of the often very favourable relational and geographical environment in which these activities are pursued, which may facilitate these endeavours. Two of our three interviewees who are at the same time micro-entrepreneurs and distributors via platforms reported having partners who "make a good living" and partly attribute the possibility of undertaking a new activity to the security this brings them. The geographical environment, for its part, should be conceived either on the micro/private level (flats in expensive, touristic neighbourhoods of Paris: Marais, Montmartre, Buttes-Chaumont), or in a broader sense (Paris). Three of our interviewees use private facilities located in the most touristic areas of Paris, some of which required a significant initial investment. Sometimes, plans to develop a freelance touristic activity go hand in hand with property investments and efforts to enhance their value, as illustrated by a dancer who recently became a host on Airbnb Experiences: "when I bought that flat, I kind of wanted to put that energy in the flat and in things, in workshops, things I could do here". Another uses a public spot at the foot of the Eiffel tower, which offers a unique and "ideal" space for her mostly international clients.

Tourism micro-entrepreneurs and platform workers thus appear to stand out from their peers working in similar capacities in different sectors, particularly in terms of level of income, level of education, but also of their experience of such working conditions. This does not lead us to formulate the hypothesis of a deliberate and emancipatory recourse to micro-entrepreneurship, as opposed to the constrained, precarious form described by Abdelnour. Indeed, the professional situations documented here are only accessible to individuals who already hold specific resources and capitals (of a social, symbolic and economic – properties, networks, etc. – nature). Is this related to the nature of the activities offered, which require self-presentation? Are the skills and know-how involved previously acquired through high-level training?

7.3   How the Platforms Select and Evaluate Skills

The platforms that connect "residents" and "tourists" offer an enchanted view of the tourist experience. The hosts who offer experiences are presented as almost ordinary "locals". On Airbnb Experiences, for instance, they are described as "passionate about their city" or "residents who will welcome you as one of their own". The emphasis on passion or on the simple condition of resident tends to downplay the work required by their activities and by the presentation and formalization of their services on offer.

At odds with this sense of an unmediated touristic experience, these individuals emphasize their specific skills, which are managed and verified by a variety of actors including the platform's salaried workers, who do not simply put them in touch with clients: there is a form of recruitment and of selection at work, both before and after the experience is conceived and completed, through evaluations and formatted procedures intended to ensure that the experience is presented and unfolds in a way that meets criteria defined by the platform (with respect for instance to descriptive blurbs, photographs, etc.). This testimony from a Brazilian artist who proposed an online experience during the pandemic reflects this process:

> OK it was a long process. It looked like… it really looked like I was applying for this, how can I say, a very big position in a company in a … I just yeah, I had to do several interviews with them. I had to give them two classes, you know, one for the people in America, one for the people in Asia.

Before hosts are put in touch with clients, there are schemes in place to help them develop their activity and structure their services, even if they are minimal (most interviewees reported that they were not directly assisted but they have access to many tools available online). One of our interviewees, who offers yoga and reiki classes on Airbnb Experiences, told us that she received a call from someone who advised her to change her pictures and description: "Basically, he told me, once my Experiences page was done, he called, said listen [...] those pics aren't right, you should have pictures of you exercising; the text is a little too long. Think about the tourist. What would you do, what would you like to read about the experience? So he did a really good job nudging me in the right direction on this." Another interviewee had to upload different pictures for her presentation, as the soft focus of some of her initial choices did not suit the platform, which requires precise descriptions in order to frame consumers' expectations and in the process prevent disappointments and negative evaluations. The choice of title for the experience was itself reformatted, which required "having sort of flashy things in the title [laughs] [...] I wouldn't have picked that title myself, for instance".

In addition to comments on design and self-presentation, interviewees also mentioned receiving advice as to their choices of dates or pricing. Platforms also holds "meetups" where hosts who offer experiences can meet and receive assistance from company representatives. The aforementioned host offering yoga classes also reported being part of a "Facebook group with community managers who are there to help us if we need help, if we have questions, if we're kind of stuck in a rut in what we do".

Also, far from being considered as "ordinary" residents, the experience providers have to prove their expertise to be listed on the platform. When they create an experience, they are asked to indicate one of three levels of proficiency: beginner-intermediate, intermediate or expert. As an interviewee noted: "if you're an expert, they might even ask for your degrees, put in the number of years of experience", although she does add that this wasn't asked of her. Conversely, some hosts willingly provide evidence of their knowledge or command of a practice by explicitly referring to their credentials in their presentation. This host who offers pastry-related activities (walks and cooking sessions) has a presentation that resembles a resumé (this is her own English translation): "I used to work for 10 years, in the chocolate family business (...) dealing with chocolates new products and marketing. I spent an amazing decade in the chocolate world. I finally follow my childhood dream and went back to school to learn Pastry in the prestigious gastronomic school FERRANDI, in Paris. I get my diploma (CAP) in pastry, and today I love teaching to my guests the professional technics to make iconic French pastries, in a fun and relax atmosphere."

The host's know-how and the quality of the experience are also evaluated after the fact. Airbnb Experiences has a team "dedicated to experience quality control, whose tasks involve reaching out to hosts whose experiences receive poor ratings to help them improve" (Léger, 2019). Consumers are likewise asked to evaluate their experience after completion by filling out a questionnaire that includes the questions: "How would you describe your host?" and "For what reasons did you think they were proficient?". Suggested answers to the former question are "They were proficient", "They had basic knowledge" and "They did not have much knowledge". For the second question, the answer choices are "The host had in-depth knowledge of their subject", "They shared unique anecdotes and were able to convey their passion", "They demonstrated a rare level of expertise or gave their own spin on the experience", or "They had very high-level knowledge or qualifications". The soft skills and personalities of the hosts are also painstakingly evaluated, in the questionnaires as well as in the customers' comments, putting indirect pressure on the formal interaction setting.

We may accordingly posit the hypothesis that these supposedly regular and autonomous service providers of sorts are actually largely supervised and subject to a full-fledged hiring process. Competences and qualifications are replaced by "expertise" and "soft skills", assessed both by the platforms and the companies' assistants, and constantly and directly by the customers. In the field of tourism, this also requires language and interculturality management skills, to better anticipate expectations.

The supervision and shaping of skills performed by the platforms is not specific to the economy of tourism; it characterizes the routine operation of organizations that provide assistance to micro-entrepreneurs in other fields. While the micro-entrepreneur status is often promoted as a means of expression of individual freedom and emancipation, in many cases self-employed workers must turn to a variety of aid structures. Some support schemes decide whether or not to back micro-entrepreneurs based on a number of criteria. As described here by an employee of ADIE (the French association for the right to economic initiative, which helps entrepreneurs raise funds), they attempt to establish "if the individual is reliable, if they can manage a budget properly. If they have experience or knowledge about the project they are setting up, if they have support from friends and relatives. First we look at the individual as a whole. And then, yes, we look at their project, see if they have good prospects...". The support provided by platforms is similar in some ways, but it also monitors forms of expression, self-presentation and marketing for consistency purposes.

7.4 From the Personal to the Professional and Vice Versa

Work via digital platforms frequently consists in the commodification of amateur and leisure practices – in fields such as crafts, as in the case of the sale handmade jewellery and other items on Etsy, or food, as with Belle Assiette's "private chef" hiring services (Jourdain & Naulin, 2019). In their research on the latter two platforms, Anne Jourdain and Sidonie Naulin note that these jobs are not necessarily performed for professionalization purposes. Often, whatever little income they generate is simply reinvested in the practice (to buy kitchenware). Still, dedicating oneself to these activities remains a means to move beyond the sphere of "intimate transactions" involving gifts and exchanges with friends and relatives, and to take part in communities of enthusiasts. Commodification thus yields forms of gratification other than simply economic ones, in part thanks to the social recognition that increased exposure brings (Jourdain & Naulin, 2019).

This blurring and overlap between work and leisure is a recurring phenomenon in studies on this kind of labour. For our interviewees, the activities they offer relate to practices they first experienced as amateurs and/or for leisure purposes. They gradually acquired a degree of expertise by committing to the practice, originally without having a professional objective in mind, with the exception of a professional dancer whose self-employed activities are an extension of her main work.

Our interviewees all report experiencing their platform work not only as a "job", something they do to earn a living, but as part of a personal lifestyle, if not a passion. This is especially the case for wellness-related experiences, as one of them puts it: "I do a lot of things, I dabble in a lot of things, and yoga... well, I went to India to attend several training sessions, but initially I was doing it for me; the point was to really learn a practice, go in-depth into what yoga is". Hence the practice is not perceived as a regular occupation, but as something that can/should be rewarding to the practitioner, as another interviewee explains: "I believe that perhaps the most important thing is the exchange, it really is, people can bring something to you, for you, so that you can pursue your life in that direction. Whereas if it's just a job, if it's to earn money then that's not rewarding. You really need to earn an energy, something that puts something of yourself in play".

The interviews reflect this reluctance to perceive platform work as a job (this applies less to persons who have developed online visits or experience during the Covid 19 pandemic for whom the goal of maintaining a revenue was important). The woman who does reiki and yoga classes says "it's not work to me; you see, I'm getting back to it now and on the contrary it pleases me so much to think that some- one is coming over; it's so great, so nice, I'm lucky, you know [...] I felt guilty about turning it into a job, making a living out of it". Sometimes this sense of individual self-fulfilment is perceived as exclusive, as per the words of a different interviewee: "Maybe it's just me, maybe I need to revisit my professional and personal sides, but it's not work, you see; it makes people feel good and it makes me feel good too".

Lastly, the porosity of the boundaries between leisure and work also raises ques- tions in terms of management of domestic space, and of the use of personal resources for professional purposes. In our interviews, we have identified different relation- ships to domestic space, which can be included in the professional practice or sepa- rated from it. In one case, in which the person works from her home, ritualized practices are used to transform her private space (a living room) into a semi-public space (by "energizing" the room and arranging things differently). Another case, in contrast, reflects an embrace of this blurring of the boundaries between domestic and work space, making a virtue out of necessity: "as far as I'm concerned, it's helped me with my project, to see sort of hybrid spaces, that kind of transform, and in which you can achieve a harmonious balance between a living space and a space for artistic practice, and I think that can be inspiring for people who are not neces- sarily experiencing this kind of interconnected thing we often have in places like this in Paris, because we don't have that much space". On the other hand, a different interviewee's trajectory reflects a growing autonomy from domestic space: whereas her first experiences took place at her home, later, as part of her entrepreneurial project, she chose to acquire premises dedicated to her professional activity, thereby drawing a line in the sand between the two. These varied uses of domestic spaces illustrate both the downplaying of work and the diffraction of the logics of work in more intimate aspects of life.

7.5   Conclusion

The early stages of this research have allowed us to draw conclusions, formulate hypotheses and identify avenues for further investigation.

First, the specific set of characteristics of the profiles we identified in our research, with a majority of educated women with previous work experience, yields insights into the debate on the democratization of work supposedly brought about by platforms. As Anne Jourdain and Sidonie Naulin (Jourdain & Naulin, 2019) have noted, the Internet is often presented as the means to overcome discrimination on the labour market. However, this work requires initial economical capital (a place to work) and cultural capital (degrees, skills, etc.). We may hypothesize that where touristic experiences in metropolises are concerned, the income and education levels of people who have recourse to micro-entrepreneurship are higher than average. This hypothesis remains to be connected to the touristic context in which these activities are performed, which could have an influence on pricing and required skills (especially language skills).

The question of the democratization of labour through the platform economy should be reformulated at a different level. While the experiences on offer are presented as coming from simple locals, the platforms actually help shape the available supply by selecting skills, self-narratives and images to highlight as indications of quality, and by offering training programmes and support services, and down the line evaluating hard and soft skills.

Here the platform's work also raises the question of the degree of autonomy of workers in their activity. The entrepreneur status is presented not as a constraint but as a choice, and the recourse to the platform is perceived as a means to gain exposure, and a source of support for communication or marketing, from which entrepreneurs do not necessarily seek to free themselves initially.

Our research also points to a hiatus between projects defined as entrepreneurial, in which individuals project themselves and invest capitals and other types of resources, and the negation of the "work" involved in these activities, experienced as extensions of a passion. This in turn raises the question of how the commodification of amateur practices and professionalization work together: what is the place of professionalization in such personal projects? In effect, whereas the fact that they constitute work is often downplayed, these practices are often major sources of income, and part of deliberate entrepreneurial projects.

Lastly, we need to consider what strategies are being implemented to manage the friction caused by the permeability of the professional and private spheres, but also to ask what advantages this situation may bring and how individuals use it for strategic purposes. The platforms themselves may incentivize or restrict the professionalization of their users/service providers, including for legal and insurance reasons (Léger, 2019). In this sense, platform work may be a genuine departure from the "work on the side" (Weber, 2009) that existed before the digital turn.

References

Abdelnour, S. (2013). L'entrepreneuriat au service des politiques sociales : la fabrique du consensus politique de l'auto-entrepreneur. *Sociétés Contemporaines, 89*, 131–154.

Abdelnour, S. (2017). *Moi, petite entreprise. Les auto-entrepreneurs, de l'utopie à la réalité.* PUF.

Abdelnour, S. (2019). Introduction. In S. Abdelnour & D. Méda (Eds.), *Les nouveaux travailleurs des applis.* PUF-Presses Universitaires de France.

Azaïs, C., Dieuaide, P., & Kesselman, D. (2017). Zone grise d'emploi, pouvoir de l'employeur et espace public: une illustration à partir du cas d'Uber. *Relations Industrielles, 72*(3), 433–456.

Boltanski, L., & Esquerre, A. (2017). *Enrichissement: une critique de la marchandise.* Gallimard.

Butler, R. (1980). The concept of a tourist area life cycle of evolution: Implications for management of resources. *Canadian Geographer / Le Géographe canadien, 24,* 5–12.

Castel, R. (2009). *La Montée des incertitudes. Travail, protections, statut de l'individu.* Éditions du Seuil.

Chareyron, G., & Jacquot, S. (2017). Airbnb ou la vie rêvée des autres. The Conversation. From https://theconversation.com/airbnb-ou-la-vie-revee-des-autres-83971

Chauvin, S. (2010). *Les agences de la précarité. Journaliers à Chicago.* Editions du Seuil

Christophe, G. (2007). Recevoir le touriste en ami. La mise en scène de l'accueil marchand en chambre d'hôtes. *Actes de la Recherche en Sciences Sociales, 5,* 14–31.

D'Amours, M., Noiseux, Y., Papinot, C., & Vallée, G. (2017). Les nouvelles frontières de la relation d'emploi. *Relations industrielles, 72*(3), 409–432.

Dardot, P., & Laval, C. (2010). *La nouvelle raison du monde, essai sur la société néolibérale.* La Découverte.

Debbage, K. G., & Ioannides, D. (1998). *The economic geography of the tourist industry, a supply-side analysis.* Routledge.

Dubois, F., & Terral, P. (2014a). Entrepreneurs passionnés. In M. Grossetti, P. P. Zalio, & P. M. Chauvin (Eds.), *Dictionnaire sociologique de l'entrepreneuriat - Introduction* (pp. 244–258). Presses de Sciences Po.

Dubois, F., & Terral, P. (2014b). La création d'entreprise dans le secteur du tourisme sportif. *Sociologies Pratiques, 1,* 53–62.

Grossetti, M., Zalio, P. P., & Chauvin, P. M. (2014). *Dictionnaire sociologique de l'entrepreneuriat - Introduction.* Presses de Sciences Po.

Guégnard, G., & Mériot, S. A. (2010). Hôtels et dépendances. *Travail et Emploi, 121,* 55–66.

Guitton, A., & Labruyère, M. (2006). *Les métiers du tourisme, approche nationale.* Net. Doc. 124.

Guitton, C., Aguettant, N., Labruyère, C., & Mahlaoui, S. (2016). *Les métiers du tourisme approche nationale.* CEREQ. *from* http://www.20.fr/publications/Net.Doc/Les-metiers-du-tourisme-approche-nationale

Ioannides, D., & Zampoukos, K. (2018). Tourism's labour geographies: Bringing tourism into work and work into tourism. *Tourism Geographies, 20*(1), 1–10.

IREST – étudiants du Master 1 Tourisme Développement et Aménagement Touristique des Territoires, (2018). L'emploi touristique à Plaine Commune (Coord. A. Condevaux, S. Jacquot), pp. 96. + pp. 269 (annexes).

IREST – étudiants du Master 1 Tourisme Développement et Aménagement Touristique des Territoires. (2019). Les nouvelles formes de travail en tourisme sur le territoire de Plaine Commune - Emplois, competences, formation (Coord. A. Condevaux, S. Jacquot, O. Kaddouri), pp.129 + pp. 206 (annexes).

Joly, M. (2020). Du «coup de foudre» à l'investissement: Les propriétaires français d'établissements touristiques dans la médina de Fès (Maroc). *Annales de Geographie, 732*(2), 53–77. Armand Colin.

Jourdain, A., & Naulin, S. (2019). Marchandiser ses loisirs sur internet : une extension du domaine du travail ? In S. Abdelnour & D. Méda (Eds.), *Les nouveaux travailleurs des applis* (pp. 47–60). PUF, la vie des idées.

Krinsky, J., & Simonet, M. (2012). Déni de travail: l'invisibilisation du travail aujourd'hui Introduction. *Sociétés Contemporaines, 3,* 5–23.

Ladkin, A. (2011). Exploring tourism labor. *Annals of Tourism Research, 38*(3), 1135–1155.

Léger, M. (2019). *L'influence des différentes perceptions de la notion d'authenticité sur le marché des activités touristiques à Paris* (master thesis, University Paris 1 Panthéon-Sorbonne).

Menger, P. M. (2014). A la recherche de l'entrepreneur. In M. Grossetti, P. P. Zalio, & P. M. Chauvin (Eds.), *Dictionnaire sociologique de l'entrepreneuriat - Introduction* (pp. 33–44). Presses de Sciences Po.

Nasom-Tissandier, H., & Sweeney, M. (2019). Les plateformes numériques de transport face au contentieux. In S. Abdelnour & D. Méda (Eds.), *Les nouveaux travailleurs des applis*. PUF.

Nogué, F. (2013). *Le tourisme, « filière d'avenir ». Développer l'emploi dans le tourisme.* Rapport commandité par le ministère du travail et de de l'emploi et le ministère du comemrce, de l'artisanat et du tourisme. From http://atout-france.fr/sites/default/files/imce/rapport_nogue.pdf

Puech, I. (2004). Le temps du remue-ménage. Conditions d'emploi et de travail de femmes de chambre. *Sociologie du travail, 46*(2), 150–167. https://doi.org/10.4000/sdt.28802

Queige, L. (2015, August). L'émergence des start-ups du secteur touristique. *Annales des Mines - Réalités industrielles, 3*, 74–76.

Réau, B. (2006). Relations amicales et rapport marchand : la relation de service dans les clubs de vacances. In J. P. Durand & M. C. Le Floch (Eds.), *La question du consentement au travail. De la servitude volontaire à l'implication contrainte* (pp. 131–140). l'Harmattan, collection Logiques sociales.

Revue Espaces. (2018). *No Entrepreneuriat et tourisme. Revue Espaces.*

Rodet, D. (2019). Capitalisme de plateforme, économie collaborative, économie solidaire : quel(s) rapport(s) ? In S. Abdelnour & D. Méda (Eds.), *Les nouveaux travailleurs des applis* (pp. 47–60). PUF, la vie des idées.

Sarfati, F., & Vivés, C. (2016). Sécuriser les intérimaires sans toucher au CDI ? La création négociée du CDI intérimaire. *La Revue de l'Ires, 88*, 3–31.

Scholz, T. (2016). *Uberworked and underpaid: How workers are disrupting the digital economy.* Polity Press. from http://pombo.free.fr/treborscholz.pdf

Shaw, G., & Williams, A. (1998). Entrepreneur ship, small business culture and tourism development. In D. Ionnides & K. Debbage (Eds.), *The economic geography of the tourist industry, a supply-side analysis* (pp. 235–255). Routledge.

Supiot, A. (2000). Les nouveaux visages de la subordination. *Droit Social, 2*, 131–145.

Torland, M. (2011). Emotional labour and job satisfaction of adventure tour leaders: Does gender matter? *Annals of Leisure Research, 14*(4), 369–389.

Veijola, S. (2010). Tourism as work. *Tourist Studies, 9*(2), 83–87.

Weber, F. (2009). *Le Travail à-côté. Une ethnographie des perceptions.* Éditions EHESS.

Weiler, B., & Black, R. (2015). *Tour guiding research. Insights, issues and implications* (p. 222). Channel View Publications.

Aurélie Condevaux is assistant professor at IREST (Institute for Research and Higher Studies on Tourism)- Paris 1 Panthéon-Sorbonne and Member of EIREST (Interdisciplinary Research Group on Tourism). Her main research interests lie in cultural performances and tourism in New Zealand and Tonga, the political aspects of heritage processes - with a particular focus on Intangible Cultural Heritage - in the Pacific and, more recently, the sharing economy and virtual tourism.

Sébastien Jacquot was elected Director of IREST (Institute for Research and Higher Studies in Tourism) at the University of Paris 1 Panthéon-Sorbonne, by the IREST Board, on July 8, 2020.

Teacher-researcher (lecturer) in geography at the University of Paris 1 Panthéon Sorbonne at IREST, since 2009. His teaching focuses on research methodology, follow-up of field workshops, courses on tourist development, heritage, tourist imaginaries, metropolisation and tourism, comparative urban policies, etc. Since 2010, he has been coordinating the Master's in tourism – Development and Territorial Tourism Planning. He has also been director of IREST since 2020, after 3 years as deputy director. As a researcher, he is a member of EA EIREST, and an associate member of UMR PRODIG. He conducts research on heritage issues, concerning both evident heritage assets (World Heritage) and what can be characterized as infra-heritage, questioning the boundaries of heritage. His work also focuses on tourism, in its links to the territorial fabric (tourist metropolisation), to memory tourism, to the digital humanities via the study of tourist social networks.

Chapter 8
The Recruitment Process in a Multinational Travel Company

Bertrand Réau

Abstract On the basis of several field studies and of data provided by the firm, this article analyzes the hiring of seasonal workers in a multinational corporation operating in the field of tourism. Because it brings face to face representatives of an institution, in a broad sense, and candidates, the job interview represents an unequal situation that tests both the legitimacy of the institution as an interrogator, and the belief of the candidates in the worthiness of the stake. The candidate temporarily accepts the authority and, sometimes, the violence of the questions, while the recruiter tries to gain access to the "truth" about the candidate, a "truth" that will make it possible to assess – according to the criteria of the institution – whether (s) he is adequate for the job. Moving away from the theories that focus exclusively on action regimes and on the forms of judgment, this article shows that the hiring process is also premised on the social dispositions and characteristics of the candidates and of the recruiters. The recruitment situation is thus a privileged moment if one wants to analyze the connivance of their respective habituses.

Keywords Seasonal employment · Recruitment · Multinational company · Habitus

As of 2007, the Leisure[1] company employed some 20,000 salaried workers – two out of three under a fixed-term work contract – in 90 different positions. Each year, it has 4000 to 6000 positions to fill, and it receives over 30,000 applications. Except for the few positions in the team that constitutes the CEO's inner circle, all senior

---

[1] In light of how some of the data used here was obtained and of the requirements of confidentiality, details pertaining to the company hereafter named Leisure have been anonymized. On this question (Roy, 2020; Goffman, 1961). This paper has been published in French in Actes de la Recherche en Sciences Sociales, n°178, July 2009.

B. Réau (✉)
LISE-CNRS-CNAM, Paris, France

© The Author(s), under exclusive license to Springer Nature Switzerland AG 2023
C. Guibert, B. Réau (eds.), *Employment and Tourism*, SpringerBriefs in Sociology, https://doi.org/10.1007/978-3-031-31659-3_8

executive jobs are filled internally.[2] Thus, external hires are almost exclusively for the lowest, minimum wage positions in resorts and at headquarters. Drawing on several field studies and on the study of corporate data (Box 8.1. Methodology), this contribution analyses the recruitment process for seasonal employees. Unlike theories focused solely on regimes of action and forms of judgment (Eymard-Duvernay & Marchal, 1997; De Larquier & Marchal, 2008), the hypothesis expounded here is that hiring conditions also depend on the dispositions and social properties of applicants and recruiters. Beyond institutional factors such as job experience and qualification, I consider whether the relative social homology between recruits and recruiters is also observed in the case of low-skilled or unskilled service employees.[3] The model that applies to the selection of executives[4] does not systematically translate to all categories of employees. Sorting and ranking operations take on variable social forms. Interrogation plays a role in this. As confrontations between representatives of an institution in the broader sense and applicants, job interviews are unequal face-to-face interactions during which the institution's legitimacy to interrogate and the applicants' belief that the game is worth it are in play. The applicant temporarily consents to submitting to authority, while the recruiter tries to discern the "truth" on the applicant, which will allow them to judge – based on the institution's criteria – whether or not they are a good fit for the position. To do this, the interviewer cross-examines various pieces of information obtained at different stages of the interrogation (resumé features, language tests, oral expression and self-presentations). For their part, the applicant tries to adjust their profile to best meet the recruiter's supposed expectations, on the uncertain basis of more or less explicit selection criteria (Hidri, 2005). The interrogation that constitutes the crux of the job interview in practice can accordingly be studied as a specific social relationship between representatives of an institution and applicants but also between individuals with more or less compatible, more or less homologous social trajectories.

---

[2] On the properties of self-taught executives (Boltanski, 1978; Pochic, 2001).

[3] This homology has been shown in the case of young engineering and business school graduates (Lazuech, 2000). This hypothesis was already advanced by Georges Benguigui (1981).

[4] There have been a number of sociological studies on the recruitment of executives. In addition to previously cited articles (Gautié et al., 2005). The hierarchical position and duties of this category of employees involve a particularly painstaking selection process. François Eymard-Duvernay and Emmanuelle Marchal note that research on this skilled staff is particularly valuable in the sense that the "evaluation methods applied [to them] have been carefully developed and tend to be diffused to other categories of employees" (Eymard-Duvernay & Marchal, 1997). The argument here is that as the hiring process for the most skilled salaried positions is the most elaborate, methodical and rigorous, its analysis helps shed light on the hiring conditions of other categories of employees.

Box 8.1   Methodology
The author conducted three field studies. The first took place in 2001, during the preparation of a PhD thesis, as an employee in two *Leisure* resorts for two six-week periods (observations, interviews, questionnaire submitted to employees (n = 244) on their social and professional trajectories and their working conditions). That same year, the author did a second study as a participant in a two-week executive training session. The third study took place in 2007 during attendance in a two-day training session on management. Additionally, for the purposes of this research, an informer carried out several observations of job interviews (at the call centre and in face-to-face settings) and conducted interviews with recruiters. He also allowed the author to collect numerous internal corporate documents (Lomba, 2008), most of which were drafted by the statistical department and the human resources department. The use of this material raised two problems. First, as these documents were collected by an employee, the author was unable to observe their construction. Then, being meant for internal use, they are so to speak incomplete for a sociologist who seeks to establish all social properties of the agents under study, unlike a human resources department. Thus, only the applicants' degrees, age, sex, and nationality were collected. As they are a core concern regarding the issue of discrimination, the hiring conditions of employees are very difficult to witness for an outside observer, who will always be suspected of performing a check (see F. Eymard-Duvernet and E. Marchal, *op. cit.*). Despite these challenges, a combination of these methods and of the analysis of various types of data yielded a few avenues of research. The jobs website of the company under study contained information on what degrees applicants were expected to hold. The list of educational institutions to which the company sends job offers gives a glimpse into the pool of potential applicants and their social properties. These data could be compared to responses to the questionnaire submitted to employees in 2001.

According to the presentation on the company's jobs website, employees enjoy unique living settings and career prospects. The first motivation for applicants is to "travel".[5] Both employees and customers are offered an "incomparable experience" in a "magnificent environment", have "rewarding social relationships" and make "discoveries". Indeed, in this company, employees and customers are expected to cultivate a very specific kind of relationship that is akin to a form of friendship between two holidaymakers rather than to servitude. Hence the hierarchy is prone

---

[5] Internal survey of applicants' motivations, Human resources department, 2007.

to minimizing differences in its presentations of resorts for employees and customers: in both cases, the objective is to sell the company. The lines are routinely
blurred: for instance, is having a drink with a customer work or not work? The
hierarchy exploit such grey areas to make employees more available. The positions
offered by *Leisure* share most of the characteristics of service jobs (Chenu, 1994;
Missègue & Cases, 2001): low-skilled, poorly paid, held mostly by people aged
under 26 and subject to intense turnover.

Unlike in many sector occupations, more men are employed at all hierarchical
levels (although proportions are reversed in one of the resorts under study). As of
2005, across all occupations, 57 per cent of employees were male (62 per cent in the
hospitality branch and 52 per cent in the leisure branch). Women are less and less
present as one moves up the hierarchical ladder: across all occupations, 35 per cent
of department heads are female (43 per cent in the leisure branch); fewer than ten
women head a resort, compared to 50 men. This inequality is particularly glaring
past the age of 27: between 21 and 26, there are roughly as many male and female
heads of department heads (ca. 1800); between 27 and 32, there are around 1200
men and 800 women; between 33 and 40, 800 men and 300 women; and above 40,
600 men and 200 women (according to human resources department figures
for 2006).

Only one in four recruits come back to work for the company during the season
or the next year. The number of positions offered actually amounts to nearly 9000
rather than 6000 due to departures (resignations and dismissals).[6] While such a turnover offers flexibility in workforce management, it does cost the company (whose
yearly revenue amounts to several billions) around half a million euros yearly. Still,
the ratio of applicants per available position is not very high: roughly one in three
applicants is hired, when departures are taken into account. Given the selection
criteria, the most skilled have strong chances of being hired. What are the consequences of this relative shortage of applicants on hiring conditions? While, for rationalization purposes, the company has invested in a recruitment software,[7] the
observation of hiring practices brings a number of paradoxes to light. The process
unfolds in two phases. The first, which is designed to prevent the system from being
overburdened,[8] consisting in sorting applicants and producing indices to rule out as
many of them as possible. The second phase involves a more delicate selection process: given the relative workforce shortage, there is a pressure to not miss the "right
candidate". In practice, applicants phone the call centre to answer a questionnaire.
If they make it through this initial selection, they are summoned for a face-to-face
interview.[9] Social proximity and distance between recruiters and applicants play a

---

[6] Human Resources department data, June 2007.

[7] This company-specific programme gives question prompts during phone interviews and records
information on applicants.

[8] These operations are performed by a variety of institutions (Benarrosh, 2000).

[9] Recruits are not guaranteed to be assigned a position: they sometimes have to wait for months to
be posted in one of the company's holiday resorts. The human resources department builds a
"reserve force" to attend to last-minute needs and fill the numerous positions that become vacant

role at each step: the recruitment process requires technical skills as well as soft skills. Even though the phone interview allows for a first assessment of articulateness, language proficiency and self-presentation in general, these serial social judgments are made most prominently during face-to-face interviews.

8.1 Remote Pre-Recruitment: Unclogging the System

The call centre is composed of seven "recruitment assistants" (six of whom are women) and a manager.[10] They sometimes speak three languages, or even four during the high season (September to May), but some are only fluent in French and intermediate in English. With the exception of Hélène, who has a master's level degree, a banker father and an engineer mother, the other recruiters are middle-class.[11] Their educational trajectories vary; some have attended mainstream higher education institutions while others hold a technical certificate (Table 8.1).

In both cases, no one has a human resources degree. The only training consists in being partnered with an assistant for a week. They are in charge of processing all mail and phone applications based on a number of criteria ranging from "skills" to "motivation", experience and language proficiency (Box 8.2. The formal steps of the pre-recruitment process). They use a programme that features some job descriptions and the language test questions. The assistants also have a job profiles booklet to help guide them in their selection. The booklet lists required skills, duties and assets per occupation. There is one for hospitality occupations (accommodation, reception, cooking) and one for leisure occupations (animation, sports, lighting, sound). The applicants are informed about the required profiles on the company's jobs website. In numerical terms, the jobs for which there are the most offers are low-skilled, often technical. For instance, to be a "polyvalent land sports animator", the applicant must be aged at least 18, have knowledge of gymnastics, be bilingual, and if possible, hold a state certificate in sport or a bachelor's degree in sport science.[12]

---

in the course of the season. For some highly technical jobs (such as cook or sound engineer), this reserve is insufficient. Generally, selected candidates who do not receive a posting often find another job before human resources gets back in touch with them. In the process, the company loses many applicants even though it has spent funds to hire them.

[10] Thirty-eight-year old Annabelle, worked 12 years as a sales representative, a shop manager, and then a choreographer in a resort. In 2001, she became an assistant at the call centre. She left the company for three years to manage a shop. In 2004, she was hired as call centre manager.

[11] Based on social background and/or educational attainment.

[12] Each employee is subject to the regulations of their country of origin, which means for instance that an Italian national can be hired without possessing a sport degree whereas a French national cannot. In a sense, these tourism operators were among the very first offshore companies. In this sector, the Leisure corporation spends the lowest proportion of its revenues on human resources (22% in 1996, compared to 35% to 60% for other similar companies), even though it touts its employees as the assets that make it stand out among the competition (Juyaux, 1996).

Box 8.2   The Formal Steps of the Pre-Recruitment Process

Whenever an applicant calls, a programme window opens on the computer to guide the assistant. While the latter does not have to follow the programme procedure to the letter, all pages must be filled out.

First, the applicant is asked whether they have already applied before, and their answer is cross-checked using their family name (the programme contains the entire history of applications).

< Never applied: the process continues.

< Already applied and programme reads "will call back": good applicant but had not yet decided to get an appointment; the process continues.

< Already applied but turned down at the call centre: if the first refusal was on grounds of lack of language proficiency or experience, the applicant is allowed to re-apply; if the reason was lack of respect or lying, a new application is denied.

< Already applied but turned down after interviewing: the applicant's file is checked to establish whether it is possible to schedule a new interview, depending on the ground for previous rejection.

< Already applied, passed the interview and has worked in a resort before: the applicant's file is checked to establish whether their season was satisfactory, in which case they may be asked in again (if they have not worked for *Leisure* in three years; otherwise the assistant simply notes that this employee wants to work again). If the former employee quit or was terminated, they are not allowed to re-apply. This can only be checked using the company's HR database – the assistants have to ask their manager to verify on their behalf.

This technical rationalization leaves room for different approaches depending on the assistant. Each can ask any question they wish to without entirely sticking to the script. The most important thing is to take as many calls as possible to maximize chances of finding "good applicants" – this is an industrial selection. Assistants cannot know everything about the wide array of technical occupations such as polyvalent technician, sound technician, lighting engineer or bookkeeper. They have to adapt to appear credible to applicants. This entails placing themselves in a position of superiority from the beginning of their interview and concealing their ignorance about the positions per se.[13] Applicants may thus be treated differently depending on who is answering the phone. For instance, Christine claims not to take age or physical appearance into consideration,[14] whereas Christian and Ingrid say this matters to them. Some recruiters do proficiency tests in languages they do not speak: in that

---

[13] Like the agents in charge of controlling the "entitled" welfare recipients studied by Vincent Dubois, they have to retain a practical control over the situation in order to perform the classification operations required by institution.

[14] CVs generally feature a photo.

Table 8.1 The Social Properties of the "Recruitment Assistants"

Recruitment assistant	Educational attainment and language proficiency	Social background	Contract	Worked at a resort
Natacha, 28, married	3 years HE, literary Arabic Bilingual French/ Arabic Intermediate English and Spanish	Blue-collar worker parents	Seasonal fixed-term + fixed-term	Yes
Christine, 27, single	3 years HE, marketing/ advertising/PR Bilingual Dutch/French Fluent English	Hasn't seen father in 20 years Mother legal corporate middle manager	Seasonal fixed-term + fixed-term	Yes
Christian, 37, single	CAP/BP vocational certificate, hairdressing French Intermediate English	Father import/export middle manager Mother unemployed	Seasonal fixed-term + fixed-term	Yes
Sonia, 40, married	2 years HE, BTS technician certificate French Intermediate English	Father sales rep Mother teacher Husband middle manager	Seasonal fixed-term + fixed-term at HQ + open-ended	Yes
Marie, 31, single	High school grad level Bilingual French/ English	Father public sector financial manager Mother secretary	Fixed-term at HQ	No
Hélène, 23, single	5 years HE, languages and corporate culture Bilingual French/ German Fluent English	Father banker Mother engineer	Fixed-term at HQ	No
Ingrid, 24, single	BEP vocational certificate, Flight attendant Bilingual French/ English	Shop owner parents	Fixed-term at HQ	No

case, they look up the questions on the programme and listen to the applicant's response. The goal of this first stage is also to filter out the most undesirable applications. Some criteria, such as being married and/or having children, are not explicit – yet they are grounds for dismissing the applicant immediately unbeknownst to them. To make sure the applicant does not find out the real reason, recruiters employ techniques designed to make them feel like they are excluding themselves:

Recruiter: *Are you free of family obligations?*

Applicant: Well, not really, I'm married with one child.

Recruiter: *And do you feel ready to leave for up to eight months abroad, far from your family. It could be Japan, Australia, Morocco… You would have to leave your wife/husband and your child behind for eight months, without a single week off.*

Applicant: Yeah, that actually won't be easy, I don't think I'm ready to be away that long and that far. Thanks for the information anyway.

Whenever the emphasis on the international aspect of the job and the length of a season fails to discourage an applicant who is married with children, the recruiter has to "get" them on something else: language proficiency, lack of experience or technical skills, even when their test results are good.

Recruiter: *Okay, so you don't have any problem without going abroad for eight months and leaving your family behind, very well, great. So now let's move on to a little English test:* [in English] *So what is your name? What job are you applying for? What do you expect from our company? Can you tell me why you want to work for our company?*

After this English test, there are two possibilities: either the assistant notifies the applicant that their language proficiency is not sufficient for the position, or the applicant speaks good English, in which case the recruiter must find fault with their proficiency in a different language, their technical skills or experience. There is also a technique to exclude applicants who are deemed insufficiently polite, hesitant, or not articulate enough. It consists in getting through part of the interview, and then letting the applicant know that their application has been registered and will be submitted to the HR manager for consideration, when in fact the assistant will send a rejection letter straight away.

If they terminate the interview prematurely, the recruiter takes the risk of having to explain themselves to the applicant. However, the evaluation of articulateness and politeness is largely subjective, and difficult to justify. By pretending that the application will be considered subsequently, they suggest that only "objective" data matter (level of qualification, professional experience, availability), thereby saving face for the institution. Whenever the assistant logs an application into the programme, they add a positive or negative comment, regardless of whether the applicant will be summoned for a later recruitment session or not. A highly unpleasant applicant can be blocked automatically by the programme if he calls back again, thanks to a kind of "black list" function that prevents very poor applicants from reapplying (Box 8.3. The phone evaluation).

Box 8.3   The Phone Evaluation

The phone interview is designed to select agents with suitable speaking skills. Articulateness is one of the company's expected standards of self-presentation, and used as an indicator of the candidate's general appearance. The evaluation of ease in communication and linguistic knowledge is likely to work against working-class applicants. This stage is also used to exclude applicants who are considered impolite or unpleasant. An applicant might respond aggressively ("I've been working for ten years, you won't be teaching me what to do"; "of course I can do that, what do you think?") and/or take the recruiter to task ("have you ever been to a resort, at least?" "[reproachfully] you have it easy asking your questions, but I for one need this job"). Likewise, an applicant who employs the familiar "tu" form of address or uses swearwords (for instance, discussing past work experience: "I got out of that gig, they were all assholes") may be considered over-familiar.

On the other hand, if the applicant has been rejected on the grounds of low language proficiency or lack of experience in the job, they are notified in a phone call, the reasons are explained and they are encouraged to reapply for the next season. In this case, the assistants do not write a record of the rejection, allowing the applicant to get a second chance later (Box 8.2. The formal steps of the pre-recruitment process).

It happens that an applicant is rejected by one of the assistants when another would have accepted them. For instance, assistants are not equally proficient in languages: Marie, who has lived in England for 12 years, speaks English far better than Christian, who has never lived in an English-speaking country and only has basic knowledge of the language. An applicant will therefore have greater chances of passing as bilingual to Christian than to Marie. As far as the applicants are concerned, many lie about their trajectory: some even go as far as to have someone else call and take their place for the interview, because they know they are bound to fail the language test. Ultimately, the assistants are recruited for their speaking skills (as well as their language level, for some of them), rather than for their degree. The same goes for the six "itinerant recruiters" in charge of the second stage of the hiring process.

8.2   The Face-to-Face Interview: Talents in the Recruiters' Image

Once they have been selected by the call centre's recruiters, applicants are summoned to one of a number of face-to-face recruitment sessions that are held in various locations in France and Northern Europe (Germany, England, Belgium, Netherlands). These sessions last a full day, with around 20 applicants for two recruiters. When applicants come in at 10 am, they are expect to hand in their completed application (the file sent to them by the assistant in charge of pre-recruitment) and submit the required documents listed in the application (certificate of employment, criminal record). The morning is devoted to a general presentation of the company. In the afternoon, each applicant is interviewed individually for twenty minutes. At the end of the day, they are included in – or excluded from – the pool of applicants per job.

Box 8.4   The Formal Individual Interview Process

- The applicant is brought in and made to feel at ease (for instance, "please sit down", "how did it go for you this morning?")
- Information on the job and the duties it entails (this is not addressed in the morning's general presentation).
- Notification of the interview's goals.

(continued)

Box 8.4 (continued)

- The applicant and his jobs: experience, duties, evaluation, career change, role-playing on a technical question.
- The applicant and the company: why choose the *Leisure* company, what are the applicant's expectations, constraints, shows, opinion on team inter-actions between activities, community life.
- The applicant as a person: qualities and flaws, attitude, impression of the interview.
- Conclusion: "Any questions?", "Anything to add?", response will be noti-fied in eight days.

Depending on the job at stake, there are variations in the questions asked dur-ing this formal process. Below, here is an example of questions asked during an interview for a bartender position:

"Hello, Miss, how are you doing?"

"So first off, is your application file complete or are you still missing a few documents?"

"How did this morning go? Was everything clear, do you have questions regarding the company presentation, the life of a resort employee? Were there things you didn't like?"

"Okay, very well, so today you're interviewing for a bartender position. What you need to know is that in our company, bartenders are not cashiers, it's only service because customers have drinks and snacks included in their package, it's *all inclusive* [in English], right? The bar product is a high-end product, we need people who are efficient and who look the part behind the bar. Usually, in resorts, there are several bars, and the bar manager has the entire team rotate between the different bars in the resort. The hours change, the locations change, there's a rotation every three weeks or every month."

"Understood? Any questions?"

"Are you currently employed? Do you have to give notice?"

"Very well, so for a start could you talk to me a little bit about your profes-sional trajectory? What would you say has been your most interesting experi-ence, and your least interesting?"

"What's your career plan? Do you have a specific goal?"

"Okay, very well, so let's do a little test, how would you go about making a Mojito? A Tequila Sunrise? A Bloody Mary?"

"Say I run into your brother in the street tomorrow, what would he tell me about you? What about your best friend?"

"What role do you play in your family?"

"What do you think you could improve in yourself? If you could change anything, what would you change?"

(continued)

Box 8.4 (continued)

"How is working for us going to be beneficial for you?"

"What did you think about this morning's group? Did you have lunch with them?"

"In terms of language, you speak English, Spanish? Okay then, let's do a little test."

"Do you have passions in life?"

"What would you take with you in your luggage?"

"When are you available?"

"How did you find out that *Leisure* was hiring?"

"Very well, miss, do you have any questions for us before you go?"

"We will get back to you within a week, alright?"

"Thank you, bye; travel home safe."

Some questions are asked for psychological assessment purposes, to gauge the applicant's ability to fit within a group ("What did you think about this morning's group?") their self-reliance ("What would you take with you in your luggage?"), but also the form and stability of their relationships and family ties (e.g., the questions on their role within the family and on their brother's or best friend's opinion of them). The recruiters are trying to establish whether the applicant will be able to adjust to a working environment that will require being far from family and friends for months on end.

When the interview is over, the itinerant recruiter must fill out a form, which is supposed to mention the position pursued by the applicant, their surname, first name, age and nationality, their availability and notice dates, the name of the recruiter, the place and date of recruitment, how the applicant learned about *Leisure*'s offer, a general assessment in scale form, and an opinion regarding criteria of evaluation of behaviour in a group during the morning's presentation. Three criteria are subject to evaluation on a scale (bad/average/good): public speaking (whether the candidate is confident, articulate, expresses coherent ideas, and has a sense of animation); creativity (in self-presentation, the audience reaction, the sense of humour); overall behaviour (participation, exchanges, enthusiasm, dynamism). The applicants' "talents" must also be listed (if they dance, play an instrument, etc.), as well as other job skills and an example of resort where the applicant could do well. To do this, recruiters have at their disposal an individual interview evaluation form (Table 8.2. Individual interview evaluation form below and Box 8.5. The documents and the language of human resources). Lastly, the recruiter must decide whether the applicant is selected. If the answer is yes, this does not yet mean that they are guaranteed a job; they will be posted (or not) depending on the company's needs, which no one mentions during the recruitment session.

Table 8.2 Individual interview evaluation form

		−	=	++
Presentation	Dress, hygiene			
Attitude	Smiles, is pleasant, polite			
Resort conditions	Housing, community life, team interactions, animation			
Adaptation skills	Adaptability, flexibility			
Team spirit/ open-mindedness	Has a few hobbies, practises sport, likes to travel, is curious			
General demeanour	Takes part in the exchanges, is congenial, driven, forward-thinking, positive			
Diploma	Has a diploma that fits the position			
Experience	Level of experience fitting the position			

Language 1: grade …/5 Language 2: grade …/5 Language 3: grade …/5 Langue 4: grade …/5.

Box 8.5 The Documents and the Language of Human Resources

The documents on which recruiters rely have been elaborated collectively by the human resources department, updating previous documents. I was unable to observe to how these documents were actually put together, but they outline recruitment criteria that were already in use by the 1960s and 1970s (Réau, 2005) with different wordings. For instance, speaking of "talents" instead of "qualification" introduces a personal dimension that naturalizes the agents' skills and by extension the agents themselves, as "talent" hints at the idea that they have innate gifts and abilities. It entails an ease in performing certain tasks that contrasts with the idea of work. The "job skills" are closer to the idea of "qualifications", but the term encompasses professional experiences, thus going beyond a definition of said skills that could otherwise be perceived as too scholastic and limiting.

In the form, all things related to appearance and bodily *hexis* come first (dress, hygiene, smiles, is pleasant, polite). On the other end, degrees and experiences come last, after living conditions, adaptation skills and hobbies and sports. They only amount to two out of nine items (including languages, which are listed separately) and not described in detail. Considering that questions pertaining to degrees and experiences have already been asked at the previous (pre-recruitment) stage, this is more of a verification/validation than a genuine evaluation (do the applicants' degrees and experiences match?). The most important criteria are probably those that pertain to individual characteristics. The evaluation system relies primarily on recruiters' perceptions of self-presentations and of the supposed adaptation skills they suggest (notwithstanding how difficult it can be, for instance, to assess an applicant's motivation).

8.3 Selling a Dream

The corporate presentation in the morning rehashes elements found on the website. The recruiters highlight the qualities of the rooms, the buffets, the food and refined products, which hardly reflects reality: in effect, employees are accommodated in two to four-person rooms and have limited access to the activities on offer (Réau, 2006). Regarding working conditions, the recruiters emphasize the "friendly" relationships between employees and managers. However, there are two very distinct categories of employees: those who deal with customers (doing reception, animation, bartending, sports) and the others (doing cleaning, cooking, technical tasks). The recruiters acknowledge that they lie to applicants: "In terms of friendliness, there really is a paradox there – we sell friendliness to employees but it only exists between employees and customers… Between employees and managers, it is virtually inexistent [and the same goes for] the employees who deal with customers and the others. The managers play them off against each other" (Valérie, an itinerant recruiter).

For instance, in mountain resorts, employees who deal with customers are slightly less paid than the others. On the other hand, they enjoy preferential access to the ski lifts. This differences in treatment create recurring problems. More generally, reception employees, for example, are expected to meet customers' expectations. In order to do so, they often have to call on those who take care of the rooms, which is a source of conflicts pertaining to the distribution and organization of work. Some managers openly take advantage of this to have their team bond by squaring off with other teams.

Regarding career development opportunities, recruiters go out of their way to point out how unique the company is: "Do you know of many companies that will present internal career development opportunities even before you're hired?". In actuality, applicants, regardless of their qualifications and experience, can only be hired at the lowest level, and they often have to wait quite long to secure an internal promotion and climb up the company ladder (Table 8.3).

The itinerant recruiters have slightly higher education levels than the call centre assistants. The only one without a higher education degree spent ten years as a resort employee and was hired specifically to recruit staff for "kids animation" (youth work) jobs. Albert and Corinne worked as a department managers in resorts for many years, whereas Vanessa and Gérald have no experience of resort work: they were hired for their degrees, as two specialists (respectively of hospitality and leisure jobs) were needed. More precisely, Albert is specialized in "family, young/ baby/mini/junior" jobs (he has ten years of resort experience in "family" jobs). Corinne is in charge of recruitment for water sports (she has a long resort experience as a water skiing manager) and hospitality jobs (she has an economics degree with a major in hospitality). Vanessa is specialized in hospitality jobs (she attended a hotel management school and has experience in the hotel and food industries). Lastly, Gérald recruits staff in sport jobs (he has a sport science degree) and management jobs (he specialized in organizational management with a major in human resources in his master's degree).

Table 8.3 The social properties of the "itinerant recruiters"

Itinerant recruiter	Education level	Social background	Contract	Resort experience
Albert, 37, single	Business high school degree + BAFA youth work certificate	Shopkeeper parents	Seasonal fixed-term then open-ended	Yes
Corinne, 31, single	Four years HE, master's in economic sciences, major in tourism/hospitality	Father engineer, stay-at-home mother	Seasonal fixed-term + fixed-term at HQ (total 8 years fixed-term)	Yes
Vanessa, 27, single	Three years HE, European hotel management degree	Father mason, mother Town clerk	Fixed-term at HQ	No
Gérald, 26, lives with partner	Five years HE, master's in management of sport organizations	Teacher parents	Fixed-term at HQ	No

Among these four recruiters, only one is strictly specialized in hospitality jobs, as a result of her training and experience. The other three received a training focused on sport and leisure during their resort experience. They have a middle-class background; to learn their job nearly all of them attended a training seminar on recruitment techniques organized by an outside firm. On the other hand, they familiarized themselves with the 90 different jobs offered by the company by sharing information. They developed their own judgment criteria on the basis of their own representations, forged in the course of their personal and professional experiences, through the questions they asked specialized colleagues and their own observations of these jobs in resorts during the induction week. The two recruiters who started in the resorts claim that the applicant's demeanour matters more to them than the job profile requirements and their technical skills. Conversely, the other two attribute far more importance to the latter. These two recruitment styles reflect different representations of what a "good employee" should be, and by extension different selection criteria.

Albert highlights the qualities of "availability", "friendliness", "kindness" and "respect", and pays little attention to "professionalism". Corinne favours "passion", "adaptation", "community life" and "generosity" – without mentioning any sort of professional aspect altogether. Those who came up working in the resorts (over seven years, in the cases of Corinne and Albert) have an approach of the "good employee" that puts far less of premium on professionalism and loyalty than their colleagues.

"If an applicant has average professional skills and great charisma, I'll take them, because that's what a 'good employee' is, it's more about your personality and people skills than being about a professional **in** your job… Anyway, [the company] doesn't know where the person's going" (Albert, itinerant recruiter).

The older recruiters have a legitimacy that rests first and foremost on their experience of the resorts. They value all the signs suggesting that applicants have "people skills". They have personalized images of the respective positions with one or several individuals that they have seen doing that job in a resort in mind. This classification is based on physical embodiments of the job (drawing on both positive and negative images of individuals that came to embody the job for them), which narrows their mental space of possibilities.

The company's new strategy requires professional, well-trained employees, with experience and mastery of several languages. They must be "friendly" and passionate about human relationships, despite a stressful working environment (15 hours a day on minimum wage under constant pressure). In that sense, the question "what is a good employee?" relates to the recruiters' perceptions of whether applicants are in effect disposed to be drudges (Pialoux, 1979). As resort work is hard and turnover is high, recruiters primarily have to find applicants for whom there is something in it, in one way or another. Albert and Corinne's criteria for instance tend to emphasize dispositions to "party" and to partake in the "animation". Observations conducted in two resorts have indeed shown that for some employees, "partying" and "moving constantly" work as derivatives that allow them to cope with difficult working conditions. These are the employees, who often possess fewer resources (in terms of education and social capital) and tend to come from more modest social backgrounds, that end up doing several seasons in the resorts. Given the virtually exclusive internal promotion system in the company, the people who decide to pursue a career in *Leisure* are precisely the ones who are least equipped to train for different jobs.

> For instance, a 33-year-old employee in charge of snorkelling wanted to work in international trade.[15] In 1993, having failed to pass his first year technician certificate in business and with no clear idea of what he wanted to do, he was elated by his experience of work in the *Leisure* company: he describes going on stage and working abroad with a foreign clientele as an "extraordinary experience". During the following winter, he trained to become a sales representative in a Parisian private school. For five to six years, he worked for *Leisure* in the summer and as a salesman in fashion in the winter. The latter job is paid on commission and occasionally brings in a "big salary", but it is subject to fluctuation. Having stopped working for *Leisure* twice, he decided to try and become a resort manager. As of 2007, he had yet to succeed.

This system creates serious relational problems between the managers, the department managers and the employees: for the latter, who often come from higher social backgrounds, are younger and more educated, it is sometimes difficult to recognize the legitimacy of hierarchical superiors who have more modest social backgrounds and are less educated. It can be reasonably assumed that this one of the causes of the recurring conflicts between hierarchical levels and of the resulting high turnover rate.[16] This high turnover rate can be understood as an indicator of bad adjustments: employees can be exceedingly or insufficiently qualified, not malleable or sociable enough, etc. The managers in place came into the company at a time when it had not

---

[15] Interview conducted in July 2001 in one of the resorts under study.

[16] The lack of internal promotion opportunities in the company also explains why it is difficult for them to retain staff over the long term.

yet launched its effort to pursue a more high-end market, and was less demanding in terms of self-presentation and quality of service. They are older, often less educated and with more modest backgrounds than the employees, and have followed an upward social trajectory thanks to the company. On the other hand, a significant proportion of the employees may have been led to expect upward social trajectories owing to their degrees. Students, in particular, find it all the more difficult to recognize the legitimacy of their managers as they are more educated than them and experience their current position as a temporary downgrading.

Gérald and Vanessa's recruitment criteria place greater emphasis on "professionalism" and "skills". They are looking for "serious" applicants who will commit to fulfilling their duties as long as possible. These are often students looking for a rewarding first professional experience; few stay for more than one season. Obviously, these profiles are not mutually exclusive and may overlap. Still, it should be noted that despite comparable diplomas and experiences,[17] the difference between hired and rejected applicants hinges on the extent to which they match the recruiter's habitus. In that sense, the interview setting is particularly important, allowing as it does to evaluate the match between applicants' and recruiters' dispositions *in situ*. Ultimately, the company's recruitment criteria are largely dependent on their appropriation by recruiters based on their own social and professional trajectories. The "truth" on the applicant comes out in the interaction between the recruiter and the recruited: appearance, self-presentation, tone and language are decisive criteria in the job interview – a particular kind of interrogation – in that they come into play in the confrontation of recruiters' and applicants' habituses.

> Despite his sport instructor diploma, an applicant who comes off as "too comfortable", speaks in a "vulgar" manner and with a *banlieue* [inner-city] accent (which does not fit the company's intended high-end image), is dismissed.[18] An applicant for a lifeguard position is ruled out because he came to the job interview wearing shorts and a tank top: the recruiters took this as evidence of sloppiness and of a perception of the company that utterly clashed with the new luxury resort policy. An applicant for a beautician position who chewed gum, wore mini-shorts, boots and a tight sweater revealing her belly button is deemed "provocative", "too sexy" and "vulgar". There is also a selection based on physical appearance involved for the most visible positions: an applicant for a hostess position, a holder of a law degree who "speaks well", is "bubbly" and "smiling" but "short and a little fat" is rejected. For those types of positions, the recruiters are looking for "tall and slender" girls with an "appealing face".[19] They know that if they send a girl who does not fit the prevailing beauty standards to work in a resort, she will be dismissed immediately upon completing her trial period. The recruiters who are also in charge of proposing assignments

---

[17] Along the same lines, Gilles Lazuech, regarding the recruitment of sales executives and engineers, asked "How can we explain that young people who graduated from the same school in the same year experience unequal struggles on the labour market?" (Lazuech, 2000).

[18] Regarding the following examples, the informant was paired with one of the aforementioned itinerant recruiters. He recounted these sessions with some embarrassment, attempting to justify his calls.

[19] The gendered evaluation of "talents" focuses on physical features: the female applicants who match the prevailing beauty standards are assumed to also have the required skills for the job. Similar assumptions are made in different settings (Guillaume & Pochic, 2007).

take this into consideration: they send girls that they perceive as "brazen" to work in adult-only party-oriented resorts and more reserved ones to family resorts. Lastly, if an applicant is pierced or wears dreadlocks, the recruiter will let them know that removing their piercing and cutting their hair are preconditions for being hired.

The recruitment of service employees is thus done by employees with comparable social backgrounds: as in the case of the executives and engineers studied by Gilles Lazuech, a social homology between recruiters and applicants can be observed (Lazuech, 2000).[20] The recruiters effectively have to act as if they were dominant in the interaction. While the *Leisure* company's strategy is based on the quality of its staff, the recruitment process is not entirely rationalized. The recruiters, some of whom are former resort employees, have sought a different working environment.[21] Their representations of what a "good employee" are based on their perception of people who run day-to-day operations in the resorts (even though they themselves have chosen to no longer work there). Being older than most applicants, they are likely aspiring to more family and professional stability than them. This mean they must pick applicants with dispositions that they have themselves deliberately left behind, but also keep in mind the company's new upmarket strategy and target applicants with a mix of qualities that is difficult to find: bilingual, skilled, articulate, artistic, presenting themselves in a way that will appeal to a wealthy clientele… and willing to work fifteen-hour days for minimum wage. Ultimately, the recruitment largely depends on the company's prior efforts of advertising to potential applicants and on the objective characteristics of the positions offered. Interrogation techniques are thus used to filter out the least eligible applicants, and to identify and convince the few "rare gems".

References

Benarrosh, Y. (2000). Tri des chômeurs : le nécessaire consensus des acteurs de l'emploi. *Travail et emploi, 81*, 9–26.

Benguigui, G. (1981). La sélection des cadres. *Sociologie du Travail, 23*, 294–307.

Boltanski, L. (1978). Les cadres autodidactes. *Actes de la Recherche en Sciences Sociales, 22*(1), 3–23.

Chenu, A. (1994). *Les Employés* (p. 66). La Découverte.

De Larquier, G., & Marchal, E. (2008). Le jugement des candidats par les entreprises lors des recrutements. *Centre d'Etudes pour l'Emploi, Document de travail, 109*.

Eymard-Duvernay, F., & Marchal, E. (1997). Façons de recruter: le jugement des compétences sur le marché du travail. *Editions Métailié, 239*.

Gautié, J., Godechot, O., & Sorignet, P. E. (2005). Arrangement institutionnel et fonctionnement du marché du travail. Le cas de la chasse de tête. *Sociologie du travail, 47*(3), 383–404.

Goffman, E. (1961). *Asylums: Essays on the condition of the social situation of mental patients and other inmates*. Anchor Books.

---

[20] G. Lazuech, op. cit.

[21] Work in resorts makes it virtually impossible to form a family with children, which leads to short careers (under 35 years) and to some extent motivates requests to be posted at headquarters.

Guillaume, C., & Pochic, S. (2007). La fabrication organisationnelle des dirigeants. Un regard sur le plafond de verre. *Travail, Genre et Sociétés, 1*, 79–103.

Hidri, O. (2005). (Trans) former son corps, stratégie d'insertion professionnelle au féminin? *Formation Emploi, 91*(1), 31–44.

Juyaux, C. (1996). Un dialogue social porteur de progrès. *ESPACES*, 27–28.

Lazuech, G. (2000). Recruter, être recrutable: l'insertion professionnelle des jeunes diplômés d'écoles d'ingénieurs et de commerce. *Formation Emploi, 69*(1), 5–19.

Lomba, C. (2008). Avant que les papiers ne rentrent dans les cartons: usages ethnographiques des documents d'entreprises. In A. M. Anne-Marie Arborio, Y. Cohen, P. Fournier, N. Hatzfeld, C. Lomba, & S. Muller (Eds.), *Observer le travail. Histoire, ethnographie, approches combinées*. La Découverte.

Missègue, N., & Cases, C. (2001). Une forte segmentation des emplois dans les activités de services. *Économie et Statistique, 344*(1), 81–108.

Pialoux, M. (1979). Jeunes sans avenir et travail intérimaire. *Actes de la Recherche en Sciences Sociales, 26*(1), 19–47.

Pochic, S. (2001). La menace du déclassement. Réflexions sur la construction et l'évolution des projets professionnels de cadres au chômage. *La Revue de l'IRES, 35*(1), 61–88.

Réau, B. (2005). *Clubs de vacances et usages sociaux du temps libre: une histoire sociale du Club Méditerranée*. Doctoral dissertation.

Réau, B. (2006). Les Devoirs de vacances: la vie quotidienne d'un Gentil Organisateur du Club Méditerranée. *Regards Sociologiques, 32*, 73–81.

Roy, D. (2020). *Un sociologue à l'usine*. La Découverte.

Bertrand Réau is a Professor at the Cnam, entitled to direct research, and holds the "Tourism and leisure travel" Chair. His recent work focuses on tourism practices and the social uses of time, the challenges of the globalisation of science and disciplinary recompositions around Studies, the relationship between tourism and ethnicity in Southeast Asia, and the development of theme parks around the world. He is notably co-author of *Sociologie du tourisme* (2016), *La sociologie de Charles Wright Mills* (2014), *Researching Elites and Power* (2020, Springer) and author of *Les Français et les vacances. Sociologie de l'offre et des pratiques de loisirs* (2011).

Chapter 9
A (Touristic) Policy Without a Ministry? A Research Note on the Effects of Training Aid During the COVID-19 Crisis

Gérard Rimbert

I do think that in France we are lucky enough to have this phenomenal diversity of territories and experiences. You can travel across France and get the sense that you're actually travelling across the world. (**Jean-Baptiste Lemoyne**, *Junior Minister for Tourism., France 24, 9 June 2020*)

Crisis is a state in which irregular things are the rule, and regular things are impossible. (**Marcel Mauss**, *In Ecrits Politiques*)

Abstract The spread of the COVID-19 pandemic has posed a challenge to the tourism sector. The actors of French tourism have seen some of their routines disrupted by this total (health, economic and logistical) crisis. Drawing on examples of touristic intelligence displayed after the 9–11 attacks, the H1N1 virus and the economic crisis of 2008–2009, institutional actors have been looking for new models, different working methods, while praising the sector for its "resilience". For their part, tourism professionals in the field have seen their activities decline or collapse altogether, as the tried-and-tested model of total engagement in the high season was replaced by a somewhat opaque range of possibilities. This research note examines the stances of the institutional actors of French tourism regarding aid to the sector at a time of crisis and the dispositions of tourism professionals to respond to them.

Keywords Covid-19 · Hospitality · Professionals · Public policies · State · Tourism

Beyond the black-and-white opposition between light-hearted optimism and uncompromising radicalism in the face of the reversal of the order of things exemplified by the two quotations above, research allows us to understand how discourses

---

G. Rimbert (✉)
LISE-CNAM, Paris, France

131
C. Guibert, B. Réau (eds.), *Employment and Tourism*, SpringerBriefs in Sociology, https://doi.org/10.1007/978-3-031-31659-3_9

on crisis and in times of crisis are the product of a longer history than that of the crisis in which they are expressed. Still, these discourses may have performative effects.

The spread of the COVID-19 pandemic has posed a challenge to the tourism sector. The actors of French tourism have seen some of their routines disrupted by this total (health, economic and logistical) crisis. Drawing on examples of touristic intelligence displayed after the 9–11 attacks, the H1N1 virus and the economic crisis of 2008–2009, institutional actors have been looking for new models, different working methods, while praising the sector for its "resilience"(Bus & Car - Tourisme de Groupe, 2020). For their part, tourism professionals in the field have seen their activities decline or collapse altogether, as the tried-and-tested model of total engagement in the high season was replaced by a somewhat opaque range of possibilities (Lainé, 2020).

This research note examines the stances of the institutional actors of French tourism regarding aid to the sector at a time of crisis and the dispositions of tourism professionals to respond to them.

Based on first- and second-hand statistics, interviews and public positions taken by actors in the sector, the analysis presented here focuses on the role of training aid schemes. It constitutes a progress report on an ongoing group research which was launched in the Spring of 2020 as the tourism sector was in the throes of the crisis,[1] and extended as part of a more formal program supported by the French Directorate for Research, Studies and Statistics (DARES).[2]

9.1   The Crisis as a Revelator of French Tourism Policy

The relevance of the Ministry of Tourism is called into question in the grey literature (Tour Hebdo, 2020). Here let us posit that a public policy is not simply the conscious, intentional product of a plenipotentiary ministry; this hypothesis is supported by the fact that over the course of its history, the sector has had to refer to the authority of several different ministries (Transportation, Culture, Foreign Affairs).

This state of affairs cannot be dismissed as the result of a bureaucratic failing to attribute a given sector to a given ministry. Indeed, attempts at unification have been regularly thwarted by centrifugal forces. For instance, the representatives of hospitality professions have established their own dialogue with public authorities, which means that to them merging with other entities for unknown results might endanger

---

[1] This study was initiated by Christophe Guibert (University of Angers, ESO), Bertrand Réau (CNAM, LISE) and Gérard Rimbert (CNAM, LISE). We are thankful to Natalie Jacquart (University of Angers) for conducting and analysis interviews and to Laure Paganelli (ASTREE) for her literature surveys.

[2] This research program has been underway since 2020 at LISE, funded by a DARES call for research projects on "The impact on the health crisis on professional skills and training".

advances they have achieved through that dialogue (a labor agreement, subsidies, etc.).

The COVID-19 crisis is an opportunity to examine what a policy without a ministry can be – i.e., a policy that does not openly present itself in the form of centralized, top-down planning, but consists in an array of principles and their translations in the field that inform the state of the sector.

9.2   A Stimulus-Driven Public Policy

The State and its institutions have redirected and reconfigured touristic flows during the economic and public health crisis. Without resorting to outright planning, since the early 2000, French authorities have attempted to anticipate the defection of foreign tourists and most importantly the need to steer French tourists towards domestic destinations.

The key actor *Atout France* (the country's tourism development agency) has for instance pursued its efforts to promote France as a destination, with the caveat that the main target audience is now made up of French nationals instead of the usual international tourists. This makes it a promotional effort of a different nature, in terms of communication talking points (with an appeal to patriotism to save the sector), of the sites being highlighted ("a change of scenery near your home" was now offered rather than exoticism) and of the style of tourism offered (vacationing outdoors to "get some fresh air" after the lockdowns and make it easier to comply with protective measures).

Atout France's May 2020 change of strategy illustrated a political will to set up a support scheme, of which that institution was to take on the marketing/demand aspect. It launched the campaigns *#CetEteJeVisiteLaFrance* and *#JeRedécouvrelaFrance* (hashtags translating roughly as "This summer I'm touring France" and "I'm rediscovering France"). With public authorities at the helm, on the supply side, the May 2000 support scheme provided economic and financial assistance (by reducing or deferring mandatory contributions). After a concerning initial assessment in September 2000, aid was offered to a wider array of businesses in October. Unsurprisingly, by the end of 2020, Atout France reported a drop in domestic revenue (−48%) that was barely less significant that the drop in international revenue (−52%) (Atout France, 2020). Although it had a middling outcome, this was a genuine tourism public policy, which was not limited to financial aid.

9.3   The Reliance on Delegation

Public authorities have indirectly influenced the sector through the forms of the financial aid offered to its different actors. Training aid schemes can be analyzed from that angle. Their manifest function is to fulfil the market's needs by professionals to meet the public's expectations and helping them improve their practices. Their latent function (Merton, 1957) is to steer financial resources towards some actors in particular rather than others (according to their eligibility and their projects' fit with current priorities).

In France, skills operators (OPCOs, *Opérateurs de Compétence*) are key actors in the provision of training schemes. In 2020, they supported tourism professionals on two urgent missions: the implementation of health measures to reopen safely after lockdowns, and the digitalization of training material (offering the added benefit of a greater ability to digitalize activities in general).

These operators have also had a center stage role in the low-key reorganization of funding flows, achieved through successive small-steps reforms:.

- 2018: fundraising reform; the operators no longer funded themselves – funds were raised for equalization purposes by France Compétences, which allocated them in function of the orientations it was entrusted with pursuing.
- 2019: The OPCOs replaced the OPCAs, and were increasingly subject to compliance with government standards.
- 2019: Businesses could receive training for funding from the government (DIRECCTE) but mostly from the OPCOs.

The center of gravity of power lies with France Compétences, which is delegated by the government, more than with the OPCOs, especially considering that the OPCOs in charge of the tourism sector are divided into four entities (AFDAS for leisure and cultural tourism, AKTO for hospitality and catering, Mobilités for transportation, and the even smaller Entreprises de Proximité (EP) department for ski areas and lifts).

Interviews with AFDAS managers revealed that the National Employment Fund (FNE) has been used both as a financial aid and as a means to shape the sector through the orientations of the training on offer and the conditions of eligibility to such training.

- 2020: Emergency training (products, digital tools, etc.) The threshold of 1.500€ per person, below which requirement levels for passing are low, boosted small-scale training programs.
- 2020: The employers best versed in bureaucratic paperwork managed to pass off routine programs as exceptional, emergency-fund eligible ones.
- 2021: The government introduced the "Collective transition" project, giving strong incentives to develop training programs aimed at facilitating career changes to account for likely job losses upon the termination of emergency financial aid schemes (such as the long-term partial activity scheme APLD).

9.4   Professional Dispositions to Perceived Aid Entitlement

A policy without a ministry would not be a policy at all if it merely amounted some hustle and bustle in ministerial palaces and their field offices without being at least a low-key presence in the everyday activities of actors on the ground.

In the case of training aid schemes, the stimulus-driven, delegation-heavy touristic policy has been rolled out first through specific rules, and then through the forms of their appropriation in the field.

9.5 A Heightened Dependence Due to the Crisis

In the process of describing the ways in which the effects of the crisis have posed a challenge to the professional robustness of the actors of tourism (in terms of adaptability, training, HR flexibility, financial strength, etc.), this research has mapped out forms of dependence of tourism professionals first on fluctuations in touristic activity, and second, on aid programs in recession periods. The term "professional robustness" is intended to refer to the actors' practices and representations here.

Surviving until they receive financial aid (and in turn making sure that that is enough to survive) is the first hurdle for tourism workers. Receiving aid also requires breaking down two barriers: those of the eligibility to aid, and of the conformity of the application.

9.5.1 Keep Going by Weathering the Storm and/or Adjusting

Seasonality and the volatility of the customer base are factors of cyclic pressure on the *tourism sector. During the 2020 health crisis, the usual peaks plummeted down to the level of* the recessive lows: the cyclic curve became a straight downward line.

This put tourism professionals in a state of heavy dependence. In troubled times, their activity is quickly fragilized, as cashflow becomes insufficient for the small operations, and in the absence of benefits for some alternative employment categories.[3]

9.5.2 Claim Eligibility

Eligibility to aid can be understood in at least two ways: first, belonging to the target category (e.g., doing cultural tours but as a wage worker, not a freelancer), and second, fitting the situation covered by the scheme (e.g., benefiting from a loan deferral but only on the basis of a given turnover threshold).

---

[3] I have deliberately refrained from using the word "precarious" here, since when business is good these statuses are not necessarily inherently precarious, considering that the good months make up for the bad ones and that lifestyles characterized by the succession of intense rush periods and slower ones have their advantages. Precarity appears in difficult times, and in some cases those are extremely rare.

This incidentally suggests a sociology of performative rankings, since the delineations of the statuses offering eligibility to aid are what "define" a tourism professional rather than a perfectly unequivocal preexisting membership to the sector.

9.5.3   Prepare Bureaucratic Applications

The aid schemes on offer, whether they are general (like the APLD and the FNE) or more specific (like the Tourism Future Support Fund or FAST, the *Terrasses éphémères* fund for outdoor dining, authorizations to open for click-and-collect orders, etc.) all require management skills and time.

Many studies (Hoggart, 1957; Siblot, 2005) on the most underprivileged members of the working classes have shown that taking steps to obtain public assistance and persevering throughout the process is not easy and requires skills. Some tourism workers find themselves in similar situations, objectively needing aid, but lacking information and familiarity with bureaucratic codes of communication.

During interviews, several tourism workers explain that they gave up on applications for aid because they felt the outcome was too uncertain, while reconstructing a sense of professional pride by recalling their past as a self-taught, self-made man or woman.

9.6   Training Aid as a Selection of the Survivors

The ongoing collective studies cited in the introduction have so far highlighted this dependence with a focus on training aid schemes. The study conducted in the Spring of 2020 shows a correlation between training dynamic and the capacity to recover, to resume activities (Fig. 9.1).

This correlation does not demonstrate the direction of causality. Does the capacity for recovery make the supply of training programs relevant, or does the training dynamic make it more likely for the business to recover? Causality can be circular, as has been shown by qualitative interviews conducted in the Spring of 2020. For instance, training staff in health protocol observance and customer conflict resolution enabled a castle in the Centre – Val-de-Loire region to reopen when the first lockdown was lifted.

A purely internal interpretation of "competences" might suggest that public authorities, through such intermediaries as France and the OPCOs, have been steering the sector towards skills reflecting the customers' demands for quality and personalized service (TravelStat & TCI Research, 2020). An external one, however, may argue that the consolidation of cross-functional skills in a sector in crisis, and which faces an even worse crisis without State assistance, could also be a policy of

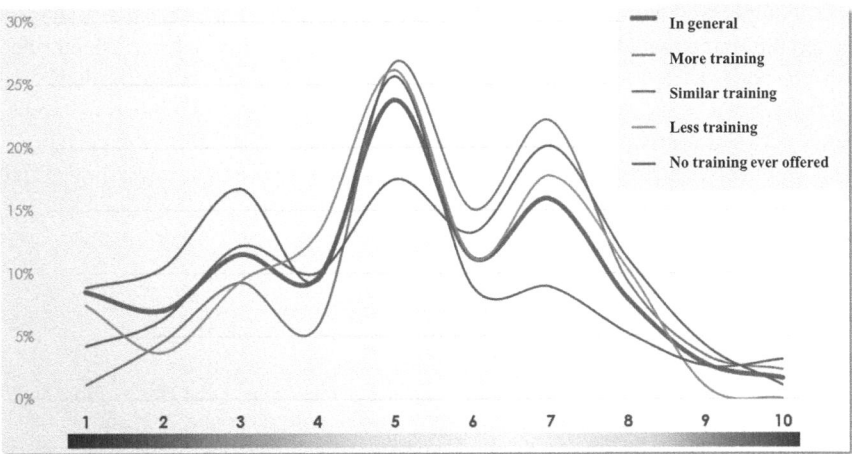

Fig. 9.1 Preparation to face recovery according to training dynamic (on a scale of 1 to 10)

employability aimed at (future) unoccupied workers that will need to be redirected towards different sectors to avoid structural unemployment. The public policy that is being implemented in the tourism sector cannot be limited to either of these options. Despite the absence of a dedicated ministry however, it is indeed a public policy, since it is possible to describe the realignment of resources and results without indulging in teleological biases.

9.7   Conclusion: State Aid as a Means to Reconfigure the Tourism Offer

This research note suggests that the metamorphosis of public policies aimed at supporting touristic activities is more of an acceleration of preexisting transformations than a genuine breakthrough. What is exceptional here is not the type of stimulus imparted by the State and its institutional partners, but the massive and brutal increase of economic situations in which tourism workers suddenly became *dependent* on aid schemes.

Even without a dedicated ministry, this policy ostensibly aimed at "defending" tourism workers effectively benefits those among the survivors endowed with the most dispositions for management. As a result, this policy impacts the style of tourism offer by relegating the most artisanal types of offer to the sidelines (ecological activists, recent converts, side businesses, etc.). Still, the support provided by training aid schemes is only one among several contributing factors to the recovery and reconfiguration of the sector, whose early trends will only become clearer after the 2021–2022 season.

All of this begs a final question: why has another political option, namely support to and through demand (in the form of vacation vouchers, for instance[4]) been ruled out? In both cases, tourism policies could still be rolled out without a dedicated ministry, but a stimulus aimed at tourists would have the merit of regulating competition between tourism offers in a way that would meet the public's expectations. This would also mean a shift from a policy of tourism in particular to a policy of free time in general (Réau, 2020).

References

Atout France. (2020). Note de conjoncture de l'Economie touristique. Atout France. From http://www.atout-france.fr/sites/default/files/imce/note_conjoncture_novembre_2020.pdf

Bus & Car – Tourisme de Groupe. (2020). Coronavirus: année noire ou année 0 d'un nouveau tourisme. Tour Hebdo. https://www.tourhebdo.com/tourismedegroupe/actualites/loisirs/coronavirus-annee-noire-ou-annee-0-dun-nouveau-tourisme-549855.php

Hoggart, R. (1957). *The uses of literacy: Aspects of working-class life*. Chatto and Windus.

Lainé, L. (2020). Voyages en Europe et dans les Dom-Tom: le grand flou perdure. *L'Echo Touristique*. *https://www.lechotouristique.com/article/voyages-en-europe-et-les-dom-tom-le-grand-flou-perdure*

Merton, R. K. (1957). *Social theory and social structure*. The Free Press.

Meynet, B. (1982). *Sur le chemin des vacances: contribution à l'élaboration d'une politique sociale des vacances, du tourisme et des loisirs*. Editions Sociales.

Réau, B. (2020). Pour une politique globale du temps libre. Les enjeux du tourisme et des loisirs en temps de crise. Institut Sapiens. From https://www.institutsapiens.fr/wp-content/uploads/2021/02/ART-les-enjeux-du-tourisme-et-des-loisirs-en-temps-de-crise.pdf

Siblot, Y. (2005). Les rapports quotidiens des classes populaires aux administrations. *Sociétés Contemporaines, 2*, 85–103.

Tour Hebdo. (2020). *Pour la création d'un grand ministère de la Culture et du Tourisme*. Tour Hebdo. https://www.tourhebdo.com/actualites/institutions/pour-la-creation-dun-grand-ministere-de-la-culture-et-du-tourisme-554681.php

TravelStat & TCI Research. (2020). Enquête de satisfaction des clientèles touristiques en France. Entreprise.Gov.fr. From https://www.entreprises.gouv.fr/files/files/tci-research-rapport-complet.pdf

Gérard Rimbert, Doctor in sociology, Gérard Rimbert has been an associate lecturer at the Cnam since the end of 2020.

---

[4] Support to demand was already part of the French Socialist Party's platform in 1981, which featured a touristic elaboration on the slogan Changer la vie [Change Life] (Meynet, 1982).

Chapter 10
Conclusion: Employment and Training in the COVID-19 Era: The Case of France

Christophe Guibert and Bertrand Réau

Abstract The case studies in this volume draw on a variety of theoretical frameworks and disciplines. Indeed, analysing employment in the tourism sector requires a plural, blended approach. Here we have fully embraced this pluridisciplinary dimension. As this book reaches its conclusion, we believe it is important to briefly discuss the exceptional period of 2020-2021. While the studies presented here predate the COVID-19 pandemic, the effects of this crisis on the economics of tourism have largely resonated with social science scholars. The years 2020 and 2021 were an entirely unprecedented, atypical historical period for tourism businesses worldwide. So far, few studies have assessed the longer-term impact of the crisis. In France, a nationwide study was commissioned by the Ministry of Labour in 2021 to capture the reality of tourism work and opportunities in the field of lifelong training. By way of a conclusion to this book, we will present the main findings of that research.

Keywords Training-employment relationship · Digital transition · Covid-19 · France

The case studies in this volume draw on a variety of theoretical frameworks and disciplines. Indeed, analysing employment in the tourism sector requires a plural, blended approach. Here we have fully embraced this pluridisciplinary dimension. As this book reaches its conclusion, we believe it is important to briefly discuss the exceptional period of 2020–2021. While the studies presented here predate the COVID-19 pandemic, the effects of this crisis on the economics of tourism have largely resonated with social science scholars. The years 2020 and 2021 were an

C. Guibert (✉)
University of Angers, Angers, France
e-mail: christophe.guibert@univ-angers.fr

B. Réau
LISE-CNRS-CNAM, Paris, France

C. Guibert, B. Réau (eds.), *Employment and Tourism*, SpringerBriefs in
Sociology, https://doi.org/10.1007/978-3-031-31659-3_10

entirely unprecedented, atypical historical period for tourism businesses worldwide. So far, few studies have assessed the longer-term impact of the crisis. In France, a nationwide study was commissioned by the Ministry of Labour in 2021 to capture the reality of tourism work and opportunities in the field of lifelong training (Guibert & Réau, 2021; Réau et al., 2021). By way of a conclusion to this book, we will present the main findings of that research.

As of 2019, according to the Ministry's data,[1] tourism was a key sector of the French economy, accounting for over 8 per cent of GDP and two million direct and indirect jobs, with an average yearly increase of 1.2 per cent (according to the Inter-Ministerial Committee for Tourism). Nine out of ten establishments were categorized as "very small businesses"; some of them had no permanent staff, relying heavily on seasonal workers, but the sector also included a few multinational hospitality corporations. The sector remained, however, difficult to delineate due to the wide variety of tourism activities and structures.

10.1   An Unprecedented Crisis in a Distinctive Sector

The impact of this crisis on the service sector, and therefore on the tourism sector, has been stronger than that of the 2008–2009 international economic crisis. Due to the diversity of the businesses that have been affected, there have been disparities in their use of government aid; some independent professions, like touristic guides, were thrown into a particularly severe depression. Depending on areas of activity and local contexts, business struggled continuously from March 2020 on (tourism in the Paris region, for instance, depends on business tourism and international travel agencies) or experienced rebounds during the summers (outdoor accommodations, regions bordering Paris, etc.). Public aid has come in the form of an emergency plan, incentives for local tourism in some regions, but also targeted training aids in the fields of the digital transition, webinar hosting and e-learning, which were sometimes made free locally. The complexity of the sector and the multiplicity of the training organizations active in France make it difficult to have a precise overall statistical breakdown on the sector. Regardless, the sector is characterized by brakes to training, and in particular to formal training through internships: there is a preference for on-the-job training following apprenticeships; employers are reluctant to part with an employee for a day or more; and possible sources of funding are difficult to identify. Public sector organizations are less affected by this.

---

[1] https://www.diplomatie.gouv.fr/fr/politique-etrangere-de-lafrance/tourisme/

10.2   The Training-Employment Relationship in the Tourism Sector

The sector is also characterized by issues surrounding the adjustment between degrees and jobs, as the academic levels of qualification of applicants increase. Professionals claim to hire more educated workers to benefit from their foreign language or digital skills; the academics in charge of initial training report that job and internship postings require qualification levels that are disproportionate to the tasks to be performed. The holders of a Diploma of Advanced Technician (BTS, a two-year degree) for tourism are generally considered to lack those skills, as well as reception and general marketing skills. However, the positions offered to those with three- or five-year degrees do not match their training or their aspirations. During the pandemic, initial training was seen as a safe haven for some students, who in this way would postpone their arrival on the job market or get there with higher levels of qualification and experience (through internships, which were hard to arrange in 2020). This mismatch between young, highly educated graduates and the expectations of businesses appears to be bound to become wider and wider.

10.3   An Acceleration of the Digital Transition and of Emerging Developments

During the first lockdown, training organizations swiftly rearranged their options to offer digital programmes; many of them had already embarked on this path, and simply accelerated the process. This has gone hand in hand with a reconsideration of the pedagogical methods to be used in remote training. The palette of digital tools has been explored in two virtually opposed directions. Some training organizations have set up complete remote training platforms, featuring community moderation, gamification or flipped learning systems. Others have favoured short webinars to reach as many learners as possible, including those who generally tend not to sign up for training due to their constraints, like family business owners—however, access to these webinars is very open, which it makes it impossible to have more specific data on the participants. Trainers on e-learning platforms have noted the diversity of learners, particularly in terms of level of qualification. The themes of these programmes have been affected by the crisis in different ways as the situation changed. During the first lockdown, emphasis was placed on immediate needs, including the implementation of health measures, remote management, how to deal with difficult customers, cancellations, and uses of digital tools to stay in touch with customers. In the hotel and food industry, trainings on takeaway services were particularly in demand when businesses reopened in the summer. Subsequently, other requests focused on anticipating a possible pick-up, for instance to deal with the new behaviours expected from tourists.

10.4   Wide Gaps in the Uses of Public Aid Schemes

While there are no global statistics for the training sector, partial data on some segments suggest that the public aid schemes offered by the French government and local authorities have been largely used. These public schemes have turned the crisis period and the time it freed up into a genuine opportunity for training, even though most of the programmes developed have tended to be short and non-certified. Still, tourism professionals have bemoaned the rigidity of these schemes and their bad fit with remote training. Additionally, these schemes have benefited the different segments in the sector to variable extents, depending especially on the size and status of businesses, which affects their capacity to take advantage of such opportunities. Those most hit by the crisis have also been those with the least resources to draw on available training opportunities. This points to the risk for these schemes to strengthen actors that were already on solid ground, as exemplified by the ability of multinational corporations to secure funds to conduct mandatory trainings as training provides. Mere financial aid proves insufficient without recommendations on programmes and advice as to their implementation.

10.5   An International Impact

Border closures, lockdowns, dramatic drops in tourist traffic and the closure of tourism service business are all factors that have impacted jobs and the economy of tourism. This is a global phenomenon: in 2020, nearly 62 million jobs were lost worldwide following the 74% drop in the number of visitors, according to the World Travel and Tourism Council (WTTC).[2]

The public aid schemes offered in France are somewhat comparable formally to those set up in other countries. Many European countries have also banked on the crisis period to develop facilitation actions in the field of training. The digital transition and domestic tourism are the most frequently emphasized facets of these funding or free training platform policies. In documents issued by international bodies, these initiatives are labelled as job protection measures, like the financial facilities and support granted to businesses. Public aid, regardless of whether it comes from the government or from local authorities, steers businesses' behaviours towards some key priorities. This raises the question of what the tourism sector has gained over the course of this historic two-year period. Has the relationship of businesses (from the very small companies to the multinational corporations) to employee training changed? Will the transformations and accelerations experienced by the tourism sector lastly change the economy and development of tourism?

---

[2] « Avec l'arrêt du tourisme, l'écosystème construit pour les voyages de masse s'est évaporé », *Le Monde*, 11 avril 2021.

The quality of tourism products rests in part on the quality of the services provided. The latter have changed to adjust to the new sanitary requirements and tourists' new aspirations. It bears reminding that very often it is the person providing the service who makes all the difference in the tourist experience. No digital platform will replace the professionals who receive tourists. Investing on the quality of tourism jobs is therefore no trifling matter: the future of the sector depends on it.

References

Guibert, C., & Réau, B. (2021). Les travailleurs du tourisme dans la tourmente. *L'Economie politique, 3*, 36–46.
Réau, B., Bureau, M. C., Guibert, C., Margaria, C., Paganelli, L., Rimbert, G., & Tuchszirer, C. (2021). *Formations et emplois en temps de crise sanitaire*. Rapport à la DARES, Programme de recherche « APR COVID. L'impact de la crise sanitaire sur les compétences et la formation professionnelle », 64.

Christophe Guibert is a sociologist, professor at the University of Angers (ESTHUA, Faculty of Tourism, Culture and Hospitality), and researcher at the "Spaces and societies" laboratory (UMR CNRS 6590). For the past 20 years, he has been examining multiple dimensions attached to tourism practices (public policies, jobs, social and cultural uses, gender, etc.) in France, but also in various foreign countries (China, Taiwan, Morocco, USA, etc.). His work is part of a dispositionalist and multi-methodological sociology. He has managed research contracts and published numerous scientific articles relating to these themes. He has published or been editor of *L'univers du surf et stratégies politiques en Aquitaine* (2006, L'Harmattan), *Tourisme et sciences sociales* (2017, L'Harmattan), *Les "sports de nature": une catégorie de l'action politique en question* (2017, Éditions du Croquant), *Emplois sportifs et saisonnalités* (2011, Logiques sociales, L'Harmattan), and *Les mondes du surf, Transformations historiques, trajectoires sociales, bifurcations technologiques* (2020, MSHA). Since 2016, he has managed two licenses and a master's degree in the field of coastal tourism in Les Sables d'Olonne, a delocalized branch of the University of Angers (France).

Bertrand Réau is a Professor at the Cnam, entitled to direct research, and holds the "Tourism and leisure travel" Chair. His recent work focuses on tourism practices and the social uses of time, the challenges of the globalization of science and disciplinary recompositions around studies, the relationship between tourism and ethnicity in Southeast Asia, and the development of theme parks around the world. He is notably co-author of *Sociologie du tourisme* (2016), *La sociologie de Charles Wright Mills* (2014), and *Researching Elites and Power* (2020, Springer) and author of *Les Français et les vacances. Sociologie de l'offre et des pratiques de loisirs* (2011).